STARS

James B. Kaler

**SCIENTIFIC
AMERICAN
LIBRARY**

A division of HPHLP
New York

Library of Congress Cataloging-in-Publication Data

Kaler, James B.
    Stars / James B. Kaler.
      p. cm.
    Includes bibliographical references and index.
    ISBN 0-7167-5033-3
    ISBN 0-7167-6031-2 (pbk)
    1. Stars. I. Title.
QB801.K25 1992
523.8—dc20

91-33125
CIP

ISSN 1040-3213

Printed in the United States of America

Scientific American Library
A division of HPHLP
New York

Distributed by W. H. Freeman and Company
41 Madison Avenue, NY, New York 10010
Houndmills, Basingstoke RG21 6XS, England

This book is number 39 of a series.

Contents

To Maxine, with love

Acknowledgments

This book has been a joy to write, a testimony to the many people who have helped and guided me along the way to its completion. Thanks to Jerry Lyons, who initiated the project and invited me to write for the Library, and to Amy Johnson, who developed the manuscript with creativity, kindness, and constant encouragement. Production deadlines were always second to quality, and she made sure I had the time I needed. Special appreciation goes to my copy editor Nancy Brooks, who loves the English language and meticulously examined my every construction. Clarity was her great goal, and I will always consider her one of my teachers. Thanks also to Elisa Adams and to Christine Hastings, who expertly managed the project through production.

This book is an amalgam of science and art. I am grateful to Travis Amos for his knowledge of the subject, his advice, and his successful searches for superior photographic images. Thanks too to illustrator Ian Worpole for the beauty and accuracy of the drawn figures and to Megan Higgins for her lovely design.

No book of this scope is possible without the aid and advice of scientific colleagues. Bruce Twarog masterfully helped rearrange parts of the text and pointed to several areas in the first draft that needed improvement. My old friend Ray White diligently made numerous corrections, and my Illinois co-workers Icko Iben and Jim Truran instructed me on the finer points of stellar theory.

I reserve the most important place for my wife, Maxine Grossman Kaler, for her support, enthusiasm, and patience. Thank you, Maxine, and thank you all.

Urbana, Illinois   August, 1991

I would like to thank the Scientific American Library and its editors, specifically Jonathan Cobb and Amy Trask, for the opportunity to update and revise *Stars* for the paperback edition. This new version incorporates several years of stellar research, and includes results from the Hubble Space Telescope (as well as several new Hubble images), from the Hipparcos parallax mission (which allows accurate distances), and from other orbiting and ground-based observatories. Minor errors have also been corrected. The result is a new and modern examination of the stellar sky.

Urbana, Illinois   August, 1998

Vincent van Gogh's "Starry Night on the Rhone."

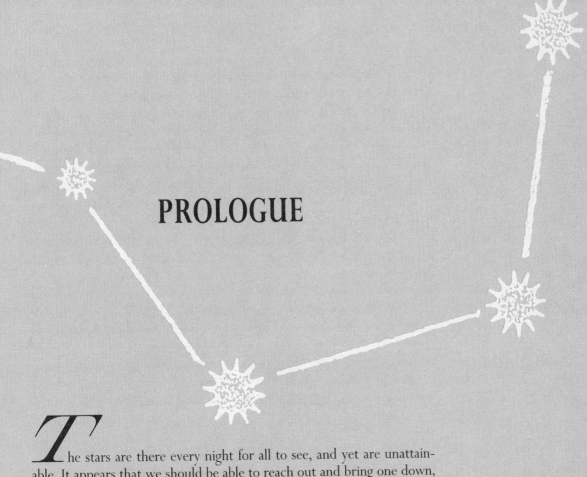

PROLOGUE

*T*he stars are there every night for all to see, and yet are unattainable. It appears that we should be able to reach out and bring one down, but they are entirely beyond our grasp. It is perhaps this dichotomy that lends the night sky its enchantment, that brings young and old out to look at the sparkling lights, and that can forge a lifetime fascination with science. It was no different for our ancestors, who associated the sky with the gods and the spirit world, but who at the same time started us on the long and difficult—but ever wonderful and scenic—road toward understanding what the stars really are, where they came from and what their fates will be, and how in truth they are so intimately associated with our own beginnings.

In your imagination go back to ancient times, and like the first astronomers begin to observe the heavens. The nighttime sky appears as a vast bowl covered with stars that fly from east to west, as the Sun does in the course of the day. Each succeeding night the spangled vault is shifted a bit westward as some stars disappear into the evening twilight and new ones emerge from the morning's fresh glow. Over a year you watch the whole assembly make one turn against the Sun, and you realize that the master of daylight is actually tracing a steady eastward path through the stars. During your vigil you also watch a bright silvery Moon change its shape and move nearly, but a dozen times more quickly, along the same track. To your increasing wonderment, five of the brightest bodies, the planets, cannot keep their places and drunkenly wander about, though also confined to a broad easterly road centered on the solar course. Other sights simply astonish you. A sudden light, a meteor, streaks across the sky—is this a star tumbling from a precarious celestial balance? Or you may see a delicate luminous ribbon, a comet, slowly work its graceful way through the nightly display. All these phenomena separate themselves out from the real stars, the thousands of lights that reliably stay their courses, that grant a sense of celestial constancy over a lifetime and link human generations over millennia.

Over the past 3000 years a steady progression of sky-watchers has sorted out the various aspects of the heavens. We have realized that the moving bodies are all close and are going around the Sun, and that our home, the Earth, is another planet to be numbered in their company. More remarkably, we have found that the real stars are astoundingly distant—hundreds of thousands, millions, *billions* of times more distant than the Sun and planets—and most importantly *that the Sun is one of them*. Stars, we have learned, are vast spheres of glowing gas powered by intense nuclear reactions deep in their cores. The Sun, typical of its cousins, is over a million kilometers across and contains the weight of nearly a million Earths. These nuclear furnaces are the principal means for the conversion of matter into energy, and are the sources and sustainers of life itself.

Collectively, the stars are the primary repository for the visible mass of the Universe, the tracers that allow us to understand the evolutionary cycle of the Galaxy, and indeed of all of creation. The serene night sky belies the churning cauldron of stellar birth and life. The stars are always coming and going, most dying quietly, some exiting in a brilliant flash of destruction, all to be replaced by others waiting their chances to be born. They are arranged in vast groups called galaxies, and in our own there are some 200 billion of them. In the part of the Universe we can see, we estimate there are over a billion galaxies, each with its stellar complement—

a hundred billion billion stars, and more beyond these, and hardly two exactly alike! They range in size from swollen mammoths that rival the solar system to tiny, incredibly dense knots of matter not much larger than a small town. Most are so feeble that they would be invisible to the naked eye even were they in the closest stellar neighborhood, while other splendid lighthouses illuminate the way across the vastness of intergalactic space.

The stars offer far more than their purely scientific or aesthetic appeal. They are natural laboratories to be used in our perpetual quest for the understanding of matter and energy. Nowhere on Earth can we make and sustain the conditions at the center of the Sun, or in the blackened chill of stellar birthplaces, or in the blistering fires of stellar death. What we learn there, we can—and have—applied here. No discipline can arise in isolation. Astronomical knowledge feeds into chemistry, physics, geology, even biology, and they pass on the favor with their own rewards. To know ourselves, we must know the stars.

The range of astronomy is vast. To some it brings to mind the Moon and the planets, and to others the billions of galaxies and the origin and destiny of the Universe itself. But when you step outside on a dark, clear night the attention is on the stars, twinkling lights that simply fill the sky, our vision encompassing fainter and fainter ones until they disappear into the blackness. So follow along as we embark on a journey that will take us from the Earth into the depths of space and on into their very hearts.

Peter Apian's plate of the northern celestial sky, from his Astronomicum Caesareum, *1540.*

Chapter One

FROM ANCIENT WONDER

STAR LORE AND THE SETTING
OF THE SKY

*I*n our modern age we focus on the *science* of astronomy, on what we have learned about the physical natures of the stars, what they are, their origins, and their fates. All too often we forget to look upward without our instruments of measurement and analysis, to gaze into the nighttime sky at the masses of stars above our heads, and to contemplate quietly their grandeur and their significance to humanity over the centuries. So if the whole tale is to be told, we must start with ancient learning that survives yet today—knowledge and lore to tie the abstract discussions that will follow to real stars in the sky, giving us a glimpse into the past and providing an unbroken connection to our forebears of so long ago.

EARTH AND SKY

To examine the stars, we need to establish their context. The heavens display a remarkable order. At a given place and date, the Sun will reliably ascend the horizon at the predicted point and time, and at 8:00 P.M. on a January evening you know without looking exactly what stars are in the sky. The movement and placement of the heavenly population can be understood with the aid of an old concept called the celestial sphere, a sphere of infinite radius that encloses the Earth at its center. It is the apparent sphere of the sky, the dome on which one sees affixed the Sun, Moon, planets, and stars; that is, we return for the moment, and for quite practical reasons, to the fiction of a geocentric universe.

We all know the Earth is a sphere, but only a few hundred years ago it was commonly believed to be flat, and various "flat-Earth societies" even now hold to that doctrine. Yet the concept of the round Earth goes back at least to Pythagoras, about 550 B.C. The evidence is legion. Ships drop away hull first as they proceed over the horizon, showing the surface to be curved. Aristotle (384–322 B.C.) discussed how stars change their positions in the sky, appearing to move north as a traveler walks south; he also observed that the outline of the shadow of the Earth, which is projected on the Moon during an eclipse, is circular. Eratosthenes of Cyrene (c. 275–195 B.C.) even measured the Earth's size! He knew that in Syene in southern Egypt the Sun was directly overhead and cast no shadow on the first day of summer; yet in Alexandria, where he lived, it was a 50th of a circle south of overhead on that date. The Earth must then have a circumference 50 times the distance between the towns. If we are right about the size of the distance unit of his time, Eratosthenes found the circumference to an accuracy of about 2 percent.

The Earth's rotation, which is responsible for making the Sun and stars appear to rise and set, is another matter. Aristotle forcibly argued that the Earth is too heavy to move, and his view held firm until the seventeenth century. Yet even in his day, some thinkers realized that the simplest explanation for the stars' march westward would be to have the Earth spin oppositely.

Now take these concepts back to their birthplace, ancient Greece, and construct the view of the sky seen by its citizens. The Earth turns on an axis speared through its center, emerging at two poles. In between the poles, splitting the Earth in half, is an equator. The angle measured perpendicular from the equator to any point is the latitude (a modern concept, abbreviated ϕ). The sky is described exactly the way it appears. You, the observer, always seem to stand on "top" of the world. Since you are apparently

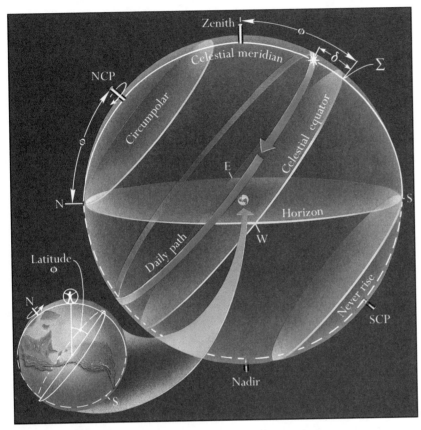

The celestial equator divides the sky into northern and southern hemispheres, and the celestial meridian, which passes from the zenith through the poles, splits the heavens into their eastern and western sides. The intersections of the horizon with the celestial meridian and equator define north, south, east, and west, and that of the equator and meridian defines the equator point, Σ. The declination of a star (δ) is the vertical angle measured to it from the celestial equator. The observer's latitude is always equal to the declination of the zenith and the altitude (the elevation angle perpendicular to the horizon) of the celestial pole. As the Earth turns from west to east, the sky seems to rotate oppositely, from east to west around the celestial poles, carrying the stars and Sun along on daily paths parallel to the celestial equator.

upright, the Earth must be tilted over in space relative to the vertical direction, with the terrestrial equator to the south by an angle equal to your latitude. The north pole is then to the north by 90° minus the latitude. If you extend a plane tangent to the Earth at your feet it will intersect the celestial sphere in a great circle (one whose center is coincident with the center of the sphere) called the horizon, which divides the sky into equally

Stars spin around the north celestial pole in this 2-hour time exposure taken at an observatory on Mauna Kea. Stars sufficiently close to the pole remain perpetually visible.

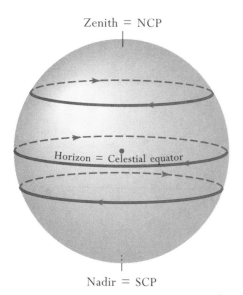

At the north pole of the Earth, the north celestial pole is in the zenith and the celestial equator is on the horizon.

sized visible and invisible hemispheres. Directly above your head is your zenith, and below your feet the invisible nadir. Your zenith and nadir are personal property; as you walk and move around the curvature of the Earth they follow along with you through the stars.

Next, extend the Earth's rotation axis and equatorial plane outward on both sides until they intersect the celestial sphere, where they respectively define the north and south celestial poles (NCP and SCP) and the celestial equator. These are the exact analogues to the terrestrial versions, the celestial equator dividing the sky into its northern and southern hemispheres. Since the Earth rotates about its geographic poles in the counterclockwise direction (as viewed from above the north pole), the observer always riding parallel to the equator, the sky will appear to counter-rotate in the opposite direction about *its* poles, all the stars appearing to move from east to west on fixed daily paths parallel to the celestial equator. Finally, the great circle that runs through the zenith, nadir, and celestial poles is called the celestial meridian, and separates the heavens into its eastern and western hemispheres. The intersections of the celestial meridian and the horizon locate north and south, and east and west are defined by the points at which the horizon meets the celestial equator. (The compass points are *not* defined by a magnetic compass, which points toward magnetic north in Canada). The celestial equator crosses the celestial meridian at the equator point, Σ.

Earth and sky are inextricably locked together. The celestial analogue to latitude on Earth is declination (δ), the angle measured perpendicularly from the celestial equator (north or south) to a star. Since the equator and poles of the sky are directly above those of the Earth, the declination of the zenith is always equal to the observer's latitude. Angles of elevation measured perpendicularly from the horizon upward to a star are called altitudes (h). The altitude of the north celestial pole is also equal to the observer's latitude. We then have some information about where we are on Earth from the orientation of the sky. Stars on the equator, those with declinations of zero degrees, rise exactly east and set exactly west. Those with northerly declinations cross the horizon to the north of these points, those with southerly declinations to the south. If the declination of a star is above a critical value, 90° minus the latitude, and thus is closer in angle to the celestial pole than the celestial pole is to the horizon, the star cannot set, and it is said to be circumpolar. Conversely, if a star's declination is farther south than the critical value, it will never rise. The result is one set of stars that is perpetually visible from a given latitude, another that is never seen.

The farther north you travel, the higher the north celestial pole, the more circumpolar stars you see, and the less you can view of the southern celestial hemisphere. If you are brave (or foolish) enough to stand on the north pole, then the NCP is in the zenith, all the stars of the northern hemisphere become circumpolar, and all of the southern are perpetually hidden. If you travel south to the equator, the poles lie right on the horizon; none of the sky is circumpolar, and you can see it *all*. Step into the southern hemisphere, and now you see the south celestial pole, and part of the northern heavens does not rise.

EARTH AND SUN

The Sun adds an additional layer of complexity. Along with the stars, it rises and sets with the Earth's rotation and traverses a daily path. However, at the same time it is always moving easterly through the stellar background about 1° per day along an annual path called the ecliptic, which is inclined to the celestial equator by an angle of 23.5°. The Sun crosses the equator moving northerly at a point called the vernal equinox on March 20, the first day of northern spring. By June 21, the first day of summer in northern latitudes, it has attained a maximum northern declination of 23.5° at the summer solstice. Then it spends six months plunging south, crossing the autumnal equinox on September 23 (the first day of fall) and by December 22 has reached the minimum declination of 23.5° south at the winter solstice (these seasonal names are reversed in the Earth's southern hemisphere, where the seasons are opposite). After 365¼ days, the Sun arrives back at its starting position. These solar reference points are fixed among the stars (at least as far as we are concerned at the moment) and have daily paths around the Earth.

The origin of the ecliptic and the solar motion is obviously the Earth's orbit around the Sun. Throughout most of human history it was widely and quite logically believed that the Sun went around the Earth. After all, if the Earth moved through space, one would expect the stars to shift back and forth annually. But as Aristotle pointed out, such stellar parallaxes were not seen. Only a lonely few, championed by Aristarchus of Samos a generation after Aristotle, pointed out the virtue of a terrestrial orbit. However, although stellar parallaxes were not detected until the nineteenth century, the theory of terrestrial revolution was finally placed on a firm footing by Copernicus in 1543 and backed up by Galileo's first telescopic observations in 1609 and 1610.

As the Earth orbits counterclockwise, the Sun will appear to move in the same direction against the distant stars. The easterly solar motion means

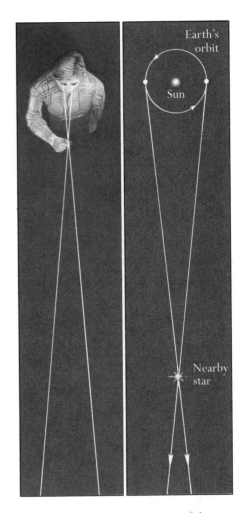

Astronomical parallax is just a version of three-dimensional vision. Look at an object with one eye, then the other; the object will seem to have moved. Similarly, as the Earth orbits the Sun, a nearby star will appear to shift back and forth against the background of distant stars.

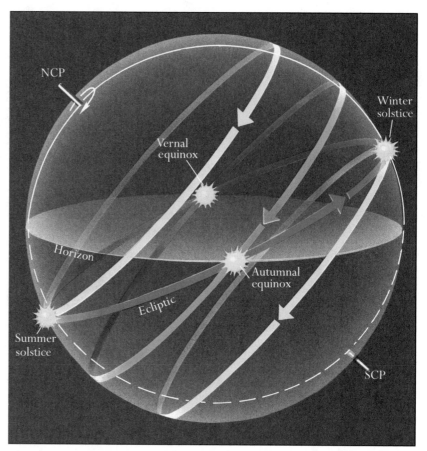

The ecliptic slants across the sky and intersects the celestial equator at the vernal and autumnal equinoxes at an angle of 23.5°. The extreme southern and northern points, the winter and summer solstices, have declinations of 23.5° south and north respectively. Each point has its own daily path around the sky, the diagram showing the vernal equinox rising. In 12 hours it will be setting, and the northern part of the ecliptic will be above the horizon. The Sun moves not quite 1° a day along the ecliptic in the direction opposite to its daily motion. The duration of daylight and the rising and setting points of the Sun depend on where it is on the ecliptic.

that the stars move westerly along their daily paths slightly faster than the Sun. Note the position of a star tonight at a specific time. Tomorrow at the same time the star will have shifted nearly 1° (360° divided by 365 days) to the west, the next night 2°. After a month it will be almost 30° farther over as the Sun slowly overtakes it. The result is that stars are seasonal in their apparitions, some visible in summer, others in winter. Our view of the heavens depends upon where the Sun is on its ecliptic, since at night we see that part of the sky in the opposite direction: When the Sun is at the vernal equinox, the autumnal and its associated stars cross the meridian at midnight and vice versa.

The north–south solar oscillation comes about because the terrestrial axis is not perpendicular to the plane of its own orbit: It is tilted, through accident of creation, by 23.5°, the so-called obliquity of the ecliptic. In northern summer, the Sun shines overhead in the northern hemisphere, and in winter overhead in the southern. The change of seasons comes about because in summer, when the Sun is most nearly overhead, the heating rays beat down most nearly perpendicularly on a patch of soil. However, in winter, the rays slant because of the low solar angle, and the same amount of sunlight is spread over a larger area, which thus cools. On the first days of spring and fall, when the Sun is at the equinoxes, declination 0°, it rises and sets exactly east and west, and days and nights are equal in length. In the summer, the high solar declination causes it to rise and set far to the north of east and west, and the days are long. Six months later, the Sun traverses the sky low to the south, and the night is king.

Since all the sky is circumpolar from the north pole, as long as the Sun is in the northern celestial hemisphere (between March 20 and September 23) it too is circumpolar. Someone riding an ice floe in the middle of the Arctic Ocean would witness six full months of daylight followed in late September and October by an ever-deepening twilight and a few months of

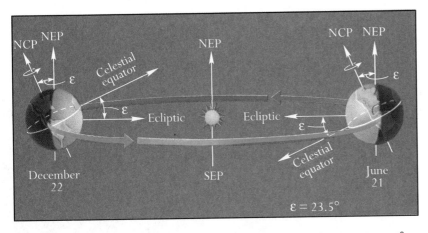

The Earth's axis is not perpendicular to the plane of its own orbit, but is tipped by 23.5°. On the left the Sun appears at the winter solstice in December, shining overhead in the southern hemisphere; on the right, the Sun has crossed the equator and is overhead in the north. The north-to-south swap is the direct—and only—cause of the change of the seasons.

absolute night. Step off the pole, and the duration of circumpolarity lessens. A limit is reached at a latitude of 66.5°N, a reference line on the globe called the Arctic Circle. There the Sun will be circumpolar only when it reaches the complementary declination of 23.5°N on June 21. Below that latitude, the Sun will rise and set throughout the year. (Technically, the line of the Arctic Circle assumes a pointlike Sun and an Earth with no atmosphere. The size of the solar disk and the lofting effect of atmospheric refraction allow the upper edge of the Sun to be circumpolar nearly a degree to the south of it.) At the Antarctic Circle, the Sun will be circumpolar on December 22, and when 24 hours of night obscure the north pole, 24 hours of sunlight illuminate the south. If the declination of the zenith equals the observer's latitude, the Sun must be overhead on June 21 and December 22 at latitudes of 23.5°N and 23.5°S respectively. These limiting parallels of latitude define the tropics, within which the Sun will annually pass overhead twice, first going north and then south.

This is the canvas of the sky. Now let the ancients paint the pictures.

THE ANCIENT CONSTELLATIONS:
STORIES IN THE SKY

With some very interesting exceptions to be examined later, the stars are randomly distributed across the heavens. Randomly, however, does not mean uniformly. They are spread quite irregularly, groups of bright ones

separated by areas with few stars at all. The eye will naturally tend to pick out patterns. A newcomer to astronomy will automatically select such groupings, or constellations, in the same way our ancestors did thousands of years ago. Already our neophyte is, as the Greeks would have it, an *astronomos,* an arranger of stars. In many cases, the configurations are so prominent that the beginner's patterns will in fact quite well match the old. As many as 8000 stars can be seen with the unaided eye over the entire celestial sphere, and the novice will rapidly discover that the simplest way to organize them for easy identification and for later study is through these groupings.

All societies, everywhere, have invented constellations. The ones with which we are now familiar in our western civilization were developed so long ago that we have lost their true origins; we know they go back to the peoples of Mesopotamia of perhaps 2000 B.C. or even earlier. These figures were adopted first by the ancient Greeks, who renamed many of them, and then by the Romans, who gave them the Latin names we use. The Arabs of the Middle Ages, preservers of classical Greek culture, then applied their own distinctive patina. The sky was becoming populated with a wonderful mixture of people, animals, and artifacts, a reflection of human history.

Forty-eight ancient constellations have been handed down to us. They were first formally recorded by the Greek mathematician Eudoxus (403–350 B.C.), and the list given its final form some 600 years later by the great Alexandrian astronomer Ptolemy in his *Syntaxis.* The constellation names commemorated myths and stories and honored gods and great heroes. People often remark on how little most (but definitely not all) of the figures look like what they are supposed to be. We should not expect them to: They are not meant to *portray,* but to *represent.*

The midnight Sun swings below the pole near Fairbanks, Alaska, on the first day of summer.

By far the most important group is the Zodiac, the set of 12 constellations that lie along the ecliptic. These were the residences not only of Apollo, the Sun god, but also of the planets that personified other Olympian gods such as the warrior Mars, the fertile Venus, and even the great king Jupiter. There are 12 because the Moon, which also traverses the Zodiac, orbits the Earth a bit over a dozen times a year, so that with every pass through its phases, the Sun occupies a different figure. (A Chinese equivalent of the Zodiac has 28 "moon stations," roughly one for each day of the lunar voyage around the Earth.)

Some of these figures are among the most familiar of the heavens. From our vantage point in Greece, Taurus glares down in late autumn and winter with a bright red eye set within his V-shaped head, really reminding us of a great bull. In Gemini you can still see the warrior twins Castor and Pollux. Leo, the lion, with his giant sickle-shaped foreparts that terminate in the bright star Regulus, regally treads the otherwise drab spring skies. And in summer, striking Scorpius, the superb red jewel Antares at its heart, awaits his prey, looking for all the world like a real scorpion basking above the southern landscape. In Homeric times Aries, the ram, an appropriate fertility symbol for the start of the growing season, was home to the vernal equinox. As the days warmed, eyes at night turned oppositely to Libra, the scales, which balanced the autumnal equinox in its pans (though at that time it was part of Scorpius). Dim Capricornus, a mysterious water-goat, held the winter solstice, and Cancer the crab, so obscure it would probably never have been an ancient constellation at all had not the path of the Sun demanded it, curled his claws around the summer. As a result, the northern terrestrial tropic, where the Sun stands overhead on June 21, is called the Tropic of Cancer, and its counterpart the Tropic of Capricorn. (Although their names have been retained, the locations of the four points have changed. The explanation awaits you below).

Of the constellations outside the band of the Zodiac, undoubtedly the most beloved of groupings are the large and small bears, Ursa Major and Ursa Minor, which stalk perpetually around the north pole. They respectively contain the Big and Little Dippers, each a seven-starred figure that almost everyone in the northern hemisphere outside big cities recognizes from childhood. The Dippers are not formal constellations but are asterisms, small prominent segments of larger figures. The Big Dipper, known in England as the Plough, is singled out in many cultures: In Arabia part of it symbolized a funeral bier, and for some Native Americans it represented six brothers and their little sister. At the end of the handle of the smaller Dipper, which is really quite hard to see as most of its stars are so faint, is Polaris, the North Star; the Big Dipper's front bowl stars always point to it.

The Big Dipper (in Ursa Major) tracks about the North Star, Polaris, followed by orange Arcturus. Between the Dippers is Draco, and part of Leo disappears to the lower left below the Black Hills of South Dakota on the northwestern horizon.

Great Orion, red Betelgeuse marking his right shoulder, contrasts smartly with blue Rigel at his left knee. From the three prominent stars of his belt, practically on the celestial equator, hangs his sword, jeweled with the Orion Nebula. At the bottom Canis Major stands on his hind legs, his snout lit by brilliant white Sirius. Canis Minor is at upper left.

In our century Polaris is less than a degree from the north celestial pole, and for those of us north of the equator, it is a signpost that allows us to find our nightly way. The small bear is circumpolar for the majority of people in the northern hemisphere, and the larger one mostly so. Behind Ursa Major is Boötes, the herdsman, with brilliant orange Arcturus—the bear-driver—ever on his tracks, always chasing, never reaching.

Then as the sky turns, and the great bear walks low beneath the pole, enters majestic Orion, placed by the gods in the winter sky opposite deadly Scorpius so that the celestial hunter need never look upon his killer. This magnificent figure so dominates northern winter heavens that the Arabs called it "the Central One." Down and to the left of Orion is his larger dog, Canis Major, which features the brightest star of the sky, Sirius, and to the upper left is the small dog, Canis Minor, lit by fair Procyon.

Sometimes several constellations considered together illustrate a familiar tale. Six figures of northern autumn concern the story of Perseus.

Proud *Cassiopeia*, the queen of King *Cepheus*, boasted that she was more beautiful than even the sea-nymphs. In anger, Neptune sent the sea-monster *Cetus* the whale to ravage Cepheus' kingdom. The king was forced to offer his daughter *Andromeda* as a sacrifice to propitiate Neptune, chaining her to a rock by the sea as fodder for Cetus. To the rescue comes *Perseus*, riding on his flying horse, *Pegasus*. Fresh from beheading the frightful gorgon Medusa, Perseus brandishes the hideous snake-haired head before Cetus, and the sea-monster is turned to stone. Celestial Cassiopeia, humiliated, spends half the year with her head downward. A lively adventure story, and an aid to remembering the placements of the stars.

The constellations and their stars revealed the dispositions of the gods and told their stories; they also held a key to worldly success. When Sirius first appeared in the morning's bright glare, the Egyptians knew the Nile would soon rise to bring them life. And listen to Hesiod, who nearly 3000 years ago advised Greek farmers:

Antinoüs, Hadrian's companion, hangs from the claws of Aquila, who carries him aloft to live with the gods. The eagle, etched in flight by Jacobo de Gehyn in Bayer's famed Uranometria *(1603), is suggested by bright Altair and the two flanking stars.*

> *At the time when the Pleiades, the daughters of Atlas, are rising,*
> *begin your harvest, and plow when they are setting.*
> *The Pleiades are hidden for forty nights and forty days,*
> *   and then, as the year reaches that point*
> *they show again, at the time you first sharpen your iron.*

THE MODERNS: FILLING THE GAPS

The classical constellations hardly fill the sky. In between many of the old figures are broad areas with few bright stars, but many faint ones, which the Greeks called the *amorphotoi,* the unformed. Furthermore, the Near Eastern and Mediterranean peoples could not see a huge portion of the far southern skies, that below a declination of about 50°S, and perforce left its stars and figures unnamed.

The new scientists of the Renaissance, however, needed better organization. They looked again at the *amorphotoi* and, aided by travelers and explorers, they went on to complete the sky. For a period of some 200 years in the seventeenth and eighteenth centuries they lavishly competed for space, inventing dozens of new constellations that reflected their own interests and discoveries.

The German astronomer Johann Hevel, or Hevelius (1611–1687), artfully filled the *amorphotoi* of the northern heavens with animals and artifacts, and to him we owe such creations as Canes Venatici, the hunting dogs south of the Big Dipper, a sextant, a lizard, a lynx, and a little lion. Around the year 1600, fancy and fanciful animals were set in the skies of the southern hemisphere by the Dutch travelers Pieter Keyser and Frederick de Houtman; these new figures are immortalized in Johannes Bayer's famous atlas, the definitive *Uranometria* of 1603. The southern population was finally completed in the mid-1700s when the Abbé Nicolas de Lacaille saluted the inventions of a blossoming technology with Microscopium, Telescopium, and Fornax (the furnace). The most famous of all the modern constellations is Crux, the Southern Cross, made from the feet of Centaurus by the seventeenth-century French astronomer Augustine Royer. Nothing dominates and represents the southern skies quite so much as this exquisite four-star figure that lies to the west of the two luminaries of the centaur. It is honored on the flags of Australia and New Zealand, and even has a "theme song"—Richard Rodgers's haunting tango "Beneath the Southern Cross," written for the television series *Victory at Sea.*

Crux, the Southern Cross.

But these represent only a fraction of the inventions. A large number (a great many very fortunately) fell by the wayside. Chief among the "defunct constellations" are those with a nationalistic bent, designed to curry favor with (and extract money from) kings and princes: Robur Carolinium, "Charles's Oak," was made from stars of Argo by no other than Edmund Halley to remind the world of Charles II, who was so impressed that he "allowed" Cambridge University to grant the great mathematician and astronomer a master's degree. The only political one that made it into the modern array is Scutum, beautifully set into the northern summer Milky Way, which represents the shield of the Polish hero and king John Sobieski, who saved Europe from the invading Turks. Others were unneeded or just plain silly. One that stands out, however, is Antinoüs, which the emperor Hadrian set below Aquila to honor his young companion, who committed suicide in the belief that his remaining years would go to his master.

The confusion of competing claims to the sky was finally straightened out at the 1922 convention of the International Astronomical Union, when the organization of professional research astronomers adopted the classical 48—actually 50, since the ship Argo was redesignated by its parts as Carina (keel), Puppis (stern), and Vela (sails)—plus 38 of the moderns, to total the currently recognized 88, all listed in Appendix 1 and Appendix 2. Subsequently, formal rectilinear constellation boundaries were also adopted in order at last to organize the heavens.

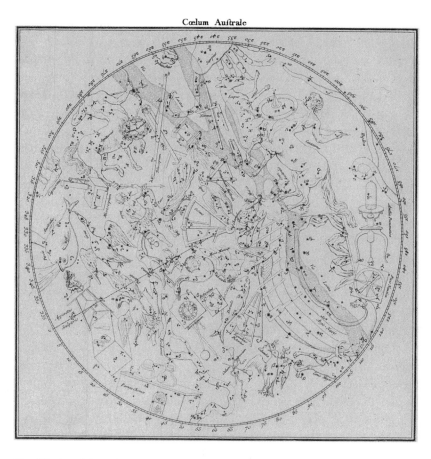

The Abbé Lacaille's south polar map displays both his and Bayer's inventions, in addition to the ancient constellations that could be seen from the latitudes of the classical civilizations.

THE MILKY WAY

One ancient image names our Galaxy itself. A sky full of stars is awesome indeed, but it is dwarfed by a great white band of light that completely encircles the heavens, poetically and aptly called the Milky Way. Like the constellations that it graces, the lore of this milky circle is ancient and deep. Its English name simply translates the Greek *galaxias,* from *gala,* "milk"— celestial milk from the breast of Hera as she nursed the vigorous, squirming infant Heracles. It was the pathway of souls to heaven, the road to paradise, a serpent, a celestial river, a sparkling cloud of dust. Imagine Galileo's astonishment when he turned his telescope onto it in 1609 to find that it

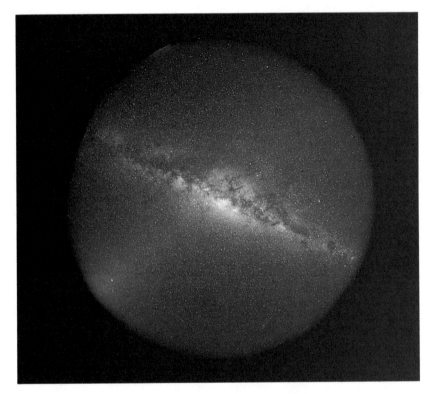

The Milky Way stretches from horizon to horizon with the center of the Galaxy above our heads; the Northern Cross is upper left, the Southern Cross lower right. The glow that emanates from the lower left is zodiacal light, produced by dust in the plane of the solar system, which scatters sunlight. The center streak marks the passage of a meteor through the atmosphere.

was actually faint stars, countless numbers of them, as we know now literally billions, the combined light of the Galaxy itself. All stars are members of galaxies of many different types. In the first analysis ours is a flat, disk-shaped system of some 200 billion stars roughly 100,000 light-years across (a light-year is the distance that light travels in a year at a speed of 300,000 km/s). If we could look at it face on we would see a beautiful system of spiral arms, best appreciated by viewing other, similar, systems.

The Milky Way is hardly uniform, but instead is broken up by enormous thick clouds of opaque interstellar dust that irregularly block starlight. The result is a band of great character, so much so that the Incas of Peru made constellations from the dark patches. Furthermore, the Sun is

Top: NGC 4565, a spiral galaxy much like ours, is seen on edge. The dust is obvious as a ridge running down the spine of the disk. Bottom: Viewed straight on, the spiral arms are apparent in M 33.

not at the center but is offset about three-fifths of the way to the edge, so that the Milky Way's visibility changes continuously: It is vastly brighter toward the galactic center in Sagittarius than it is in the opposite direction toward Taurus and Auriga. On summer evenings in the northern hemisphere the Milky Way streams out of the northeast from Cassiopeia and overhead through Cygnus, where a dust cloud, the Great Rift, splits it in two. The western branch descends through Ophiuchus and into Scorpius, where it produces an extraordinary cascade of stars; the eastern branch plunges even more brightly through Aquila and Scutum (where Sobieski's shield is made by an intense star cloud) on into the galactic core in Sagittarius, where it slips below the southern horizon. In northern winter, by contrast, we can hardly see the Milky Way in Auriga, and it brightens only some as it passes south through Canis Major, giving but a hint below the dog of the glories hidden by the Earth. The true majesty of the Milky Way is reserved for the lucky dwellers of the southern hemisphere, where Sagittarius arches his bow overhead, and spectacular cataracts of stars tumble down the heavens to either side, a singular sight that no one who sees it could ever forget.

We who live in the cities and towns forget this magnificent display of nature, one that no photograph, however skillfully or artistically made, can possibly match. Go out into the country, dozens of miles away from the lights, into the mountains if need be, to see for yourself.

THE NAMING OF STARS

The constellations parcel the celestial vault into convenient segments. Now let us narrow our view to individual stars. For any kind of study, from simple sky-watching to astrophysical research, they must be identified. A principal qualifier is their brightness, ranging from brilliant Sirius, which can be seen through a window of a modestly lit room, to as faint as you can see by eye and by telescope. Before anything else we must have a method of measurement. Over 2000 years ago, as part of the best star catalogue of the time, the Greek astronomer Hipparchus divided stellar brightnesses into six divisions called magnitudes, the first representing the most luminous stars and the sixth the faintest that could be seen. Polaris and most of the stars of the Big Dipper are second magnitude, the stars that flank Altair and represent the eagle are third, the handle stars of the Little Dipper are fourth, and the faintest star of its bowl fifth. A truly dark, moonless night away from city lights is needed to see fifth and sixth. The fainter the category, the more stars there are in it, and when the weakest come into view, the sky becomes wonderfully filled.

Nineteenth-century astronomers found that Hipparchus' scale was logarithmic in nature, and that the first-magnitude stars were as a class about 100 times brighter than those of the sixth; the obvious step was to make the ratio exact by quantifying the system. If five successive changes in magnitude correspond to a factor in brightness of 100, then each magnitude division increases or decreases the brightness by the fifth root of 100, or 2.512 . . . ; second magnitude, for example, is 2.5 times brighter than third. To establish the scale, the astronomers selected a set of faint stars near the sky's north pole, around Polaris, and fixed their average magnitude at 6.00 (the general term "sixth magnitude" stretches from 5.51 to 6.50 and so on). Twenty-two stars are brighter than magnitude 1.5. Although they are loosely called "first magnitude," their range is so large that eight are of zeroth magnitude (between -0.49 and $+0.50$), and two, Canopus in Carina (Argo) and Sirius, are even of the minus first. The nearby planets are even brighter: Jupiter and Mars reach -3, and Venus is sufficiently radiant at -5 to be seen in broad daylight with the naked eye. Continuing the scale, the full Moon, under which you cannot quite read, is -12, and the Sun, which will destroy the eye, is -27. Telescopically, of course, we can see fainter than sixth, and the largest instruments are capable of recording stars down to the 29th magnitude, 23 magnitudes less luminous than the eye can see alone, an astonishing factor of 6×10^8. To the physicist the system may seem bizarre and needlessly complicated, but it works well in a celestial context and one rapidly gets used to it.

Now we assign names. About 2000 naked-eye stars still carry their ancient proper names, adopted from a variety of cultures and languages (the flavor of them is given in the list of the brightest 40 in Appendix 3, all those down to and including Polaris). A few derive from Greek, like Sirius, which means "scorching," an obvious reference to its brilliance, and Procyon, from *pro kuon,* "before the dog" (it rises ahead of the dog-star Sirius). A few more, like Capella, "she-goat," and Spica, "ear of wheat," are Latin. The overwhelming majority, however, are of Arabic extraction. Much of Greek astronomy was translated into Arabic before it returned to Europe in the Middle Ages. The Arabs embraced the Greek images at the expense of their own, but named the stars in their language largely according to their positions within the Greek figures. (We have inherited Greek constellations with Latin names made up of Arabic stars—a wonderful amalgam!) Thus Deneb, which means "tail," pops up in Cygnus at the end of the swan and also appears as Denebola, in the hind quarters of Leo, as Deneb Kaitos, the tail of Cetus the whale, and Deneb Algedi in Capricornus. Occasionally, the original Arabic figure shines through. Alkaid in Ursa Major refers to the "chief of mourners" because the Arabic image of this constellation was not a bear but a bier. Most of these names would be quite unintelligible to a modern Arab, as they were highly distorted and corrupted upon retranslation into the European languages. Betelgeuse, originally Bet al Jauza, "the right hand of the Central One," was mangled and contracted and finally retranslated as the "armpit" of the giant we know as Orion.

The proper names, however, are neither unique nor easy to remember. Bayer solved the problem in his *Uranometria* by assigning Greek-letter names to stars within a constellation more or less in order of stellar brightness and combining the letter with the genitive form of the figure's name: Betelgeuse is thus "α of Orion," or "α Orionis." Another popular method originated with the English astronomer John Flamsteed (1646–1719), who created the best celestial maps of the time. He ordered all the naked-eye stars by position from west to east within the constellations, and a short time later the Frenchman Joseph Lalande applied numbers to them. By this system Betelgeuse is also 58 Orionis. These Flamsteed numbers are generally used for the stars that do not carry Greek-letter names; proper names are usually employed only within the first magnitude. Commonly, just the three-letter abbreviation of the constellation name is used, so Betelgeuse is simply α Ori, and 61 Cygni, the first star whose distance from Earth was found, is 61 Cyg.

For the telescopic stars, which require designations as well, we resort to various catalogue numbering systems. In all, an extraordinary 15 million stars carry some kind of formal label, yet these are but a tiny fraction of the billions of stars that can actually be observed. Even today, the vast majority of stars are anonymous points of light. However, with various maps and atlases the named ones are sufficient to find our way easily and accurately around the sky—you can begin with the star maps in Appendix 4.

THE MEASUREMENT OF THE SKY
AND THE EARTH

Our vision has passed from the wide-open sky with its constellations and brilliant Milky Way to the individual stars, and now we must narrow our focus even more. In a report of a bright naked-eye comet or a brilliant exploding star, it might be sufficient simply to say that it is in Gemini or

Delphinus. However, that information would not help much to locate a faint telescopic star, one among the millions visible in any Milky Way photograph. We must do something much more precise, and establish ways of actually determining positions in the sky. Such a scheme is the only means by which astronomers can begin to watch the stars' movements relative to one another, that is, to establish the dynamics of the Galaxy itself. Moreover, it allows us a way of measuring the flow of time and of navigating the terrestrial globe.

Latitude and declination are insufficient by themselves to locate a city on the two-dimensional surface of the globe or a star on the celestial sphere. Consider the Earth. A meridian is a circle on the globe perpendicular to the equator passing through a given point and the poles. The prime meridian, set by historical convention rather than scientific necessity, passes through the site of the old Royal Greenwich Observatory in England, and defines a zero point where it crosses the equator. Now select a city, pass another meridian through it, and note where this meridian intersects the equator. Longitude (λ) is measured between the meridians, always along the equator, east or west of Greenwich. Latitude (ϕ) is gauged along a meridian north or south of the equator.

Next, transfer this basic concept to the sky. This time, pass what is called an hour circle from the NCP through the star to the SCP and see where it crosses the celestial equator. The arc measured from the equator point, Σ, westward (all the way around if need be) to this point of intersection is the hour angle (HA), and the arc measured along the hour circle from the celestial equator to the star is the declination. Hour angle is commonly measured in clock units, in which the circle is divided into 24 equal parts called hours, each of which must be equal to 15° (that is, 360/24). Now we can measure the turn of the sky with a clock, as the hour angle increases uniformly with time. Each hour is divided into 60 minutes (m), so that 1° must equal 4^m of time, that is, the sky turns 1° every 4 minutes: Go out and watch for yourself. These minutes of time must be distinguished from "minutes of arc," the 60 subdivisions of the degree, which are denoted by a single quotation mark. If 1° is 4^m, then $1^m = 15'$ (and likewise, 1 second of time equals 15 seconds of arc).

Since daily paths are parallel to the celestial equator, declinations are constant with time. But hour angles are always changing. What we need is a zero point on the equator that turns *with* the stars. The naïve course would be to proceed as we do on Earth, and pick a favorite star to define a "prime hour circle" as Greenwich defines the prime meridian. However, that will not work at all because, unlike Greenwich, the stars move (well, yes, there is continental drift, but it is so slow as to have no significant

The prime meridian at Greenwich.

effect). If we pegged a coordinate system to Vega, for example, it would serve only as a standard for measurements of motion relative to itself. We need a zero that is divorced from the stars altogether, and for that we use the vernal equinox, which can be defined by observations of the Sun. We measure right ascension, α, from the vernal equinox eastward (opposite to hour angle) to the intersection of a star's hour circle and the celestial

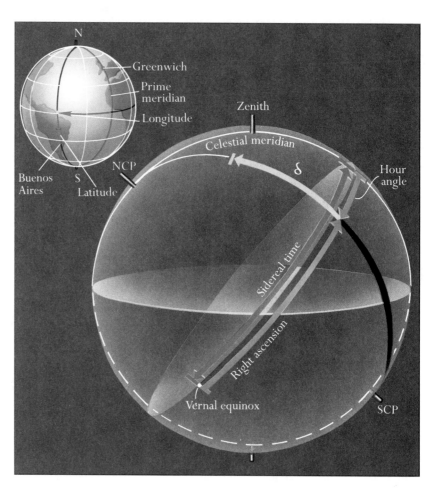

On the Earth (above, left) meridians are drawn between the poles perpendicular to the equator. Longitude is the arc along the equator between a selected meridian and the prime meridian through Greenwich, England, near London. The sky has hour circles drawn between the celestial poles through a star. Hour angle is the arc on the equator between the meridian and the hour circle. Right ascension is the arc measured in the other direction from the vernal equinox to the hour circle. Sidereal time is the hour angle of the vernal equinox, and it is always equal to the hour angle of a star plus its right ascension.

equator. This arc coupled with declination, δ, finishes the job. Until the mid-1980s, α and δ were measured for about a million stars. Because of the need for accurate guides for the *Hubble Space Telescope,* that number has now risen to an astounding 15 million. The sky is known very well indeed.

These measurements allow us to find our way in the world. Look first at time, which in its simplest definition is just the hour angle of the Sun. At

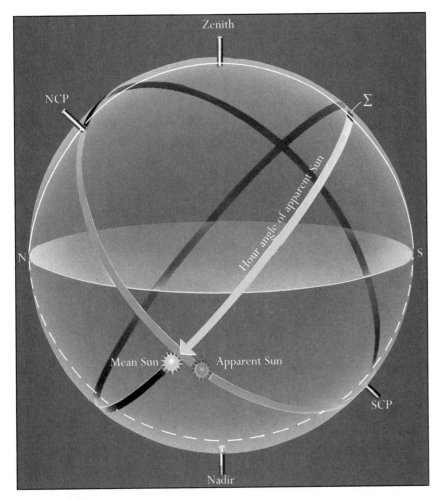

Solar time is given by the hour angle of the Sun plus 12 hours. Here the Sun is shown on February 15 (when it is south of the equator). Its hour angle is 8^h, so the time is $8 + 12 = 20^h$, or 8:00 P.M. The apparent Sun travels at a variable rate on the ecliptic and is slightly out of synchrony with the mean Sun, which moves at a constant rate along the celestial equator. On February 15, the mean Sun has fallen 3.5° behind the apparent Sun. The hour angle of the mean Sun is then 8^h 14^m, and the local mean solar time is 20^h 14^m or 8:14 P.M.

noon, when the Sun crosses the celestial meridian in daylight, the hour angle is 0^h, and when it crosses below the pole at night, it is 12^h. Since by convention the day starts at midnight, 12^h is just added to all the hour angles, making noon 12^h and midnight 24^h or 0^h. This local apparent solar time, LAST, can be read on a simple sundial.

The real (or apparent) Sun, however, is a poor timekeeper. The Earth's orbit is elliptical, and so the Sun must move at a variable rate; moreover, the Sun moves on the tilted ecliptic, but hour angles are measured on the equator. The result is that the interval between successive passages of the Sun across the meridian is not constant, and some days are slightly longer than others. We overcome this problem by inventing a *mean* Sun that steadily treads the celestial equator, keeping an average pace with the real body, and define a local mean solar time, LMST, as the hour angle of the mean Sun plus 12 hours. The difference, the equation of time, is easily calculated, so that a sundial can be readily corrected.

A friend standing 30 meters directly east of you will see the Sun shifted to the west and its hour angle increased by about a second of arc. If you compare watches, your friend's will read a 15th of a second later. Local time, then, depends on longitude, every degree corresponding to 4 minutes of time and every hour to 15°. That means every town in every state or province will keep a different time on its clocks; imagine a railroad, airline, or television schedule! The solution is to standardize the time by dividing the Earth into 24 sectors centered on meridians 15° apart. In principle,

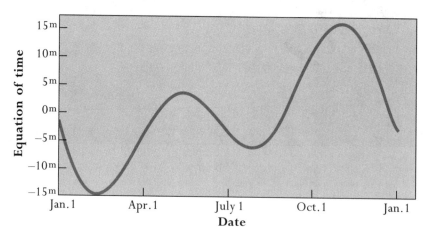

The equation of time, plotted against the date, is the difference between local apparent solar time and local mean solar time.

The photographer took a picture of the Sun every 10 days throughout the year at the same mean solar time. We see the Sun moving back and forth across the equator between declinations 23.5° north and south; it also moves east and west—ahead of and behind its average position—creating the analemma: a figure often seen on terrestrial globes.

everyone within 7.5° of a "standard" meridian would keep its time: For example, Eastern Standard Time refers to 75°W, Central European Time to 15°E, and so on. (In practice, however, the time-zone boundaries can be wildly distorted by social and political considerations.) Standard time is what you keep on your kitchen clock. We can also define a worldwide standard called Universal Time, UT, as the local mean solar time at the Greenwich meridian. Conversely, we measure longitude by time differences; we just compare the local clock to the UT, subtract the two, and convert to degrees. No clocks, no longitude—one reason it was so dangerous to try to navigate the globe before the first good seagoing clocks were invented in the mid-eighteenth century.

With the altitude of the pole and the time difference between your location and Greenwich you know both your latitude and longitude, and thus the position of your ship or town. But there is a simpler way. Both terrestrial coordinates can be determined simultaneously with such a simple instrument as the hand-held sextant. The altitudes of stars depend upon the latitude, the longitude, and the time of day. If we know the time, we can fix

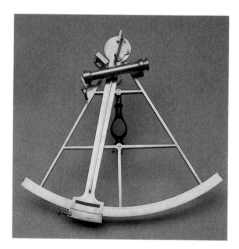

A sextant of the kind that was used by Captain James Cook in his remarkable explorations of the Earth in the mid-1700s. The navigator looks through the small telescope at the horizon. A system of two mirrors, one mounted on a movable arm, allows the observer to superimpose the image of a star onto the horizon. The star's altitude is then read from the position of the movable mirror, which carries a pointer on a graduated arc. Three altitudes and the Universal Time allow the calculation of latitude and longitude. Modern sextants look very much the same. Navigation is now accomplished with amazing precision by observing the Earth satellites of the Global Positioning System, but the old-fashioned way must still be learned as a backup.

our position by measuring the altitudes of only three stars. That is, there is only one place on Earth where those three stars will have those altitudes at that particular time. Just a little bit of trigonometry will solve the problem, and the globe is ours to roam, a gift from the sky.

Solar time does not help to locate the stars, however. For that, we need "star" or sidereal time, which is defined as the hour angle of the vernal equinox. Imagine the mean Sun to be on the meridian on September 21, when it is coincident with the autumnal equinox: The local mean solar time is 12 hours, or noon. The vernal equinox is opposite, has an hour angle of 12^h, and the local sidereal time is 12 hours as well—that is, the two kinds of time give the same value.

Now as the Earth turns and the sky rotates, the mean Sun will appear to move to the east of the equinox. The next day, September 22, the equinox will return to the meridian *before* the Sun. Since the Sun has moved nearly 1° east it will take it another 4 minutes (actually $3^m 56.56^s$) to return, by which moment the sidereal time will be $12^h 04^m$. We see two intimately related things. First, the day according to the stars is not quite 4 minutes *shorter* than that reckoned by the Sun, and sidereal time advances at a rate of nearly 4 minutes per day over solar time. It is the sidereal day and time that are fundamental; that is what an outside observer, a hypothetical Martian, would see. Second, over the course of the year, the sidereal clock gains a full day over the solar, rendering 366¼ sidereal days in the year.

Because the hour angle of the vernal equinox must be exactly equal to the right ascension of the meridian, we can find the LST, local sidereal time, by noting the right ascensions of stars as they cross it. If we know the date—actually the elapsed time between the last passage of the mean Sun across the autumnal equinox—and the time of the observation, we can calculate the amount that the sidereal clock has gained over the solar, and consequently can reckon the LMST. We then need only apply a correction for longitude to arrive at standard or Universal Time. In actual practice, however, the stars are no longer used. Sidereal time is determined by observing distant quasars—powerful pointlike beacons that lie among the galaxies at the distant edge of the Universe—with radio telescopes, which provide considerably more accuracy.

Conversely, the astronomer can determine the right ascensions of stars by timing their passages across the meridian with a sidereal clock. And if we know the precise latitude, the declinations are easily found from the observed altitudes above the horizon. (No circular reasoning is involved, since in practice LST and α are found simultaneously by iterative methods.)

Since hour angle and right ascension are respectively measured in opposite directions from the meridian and the vernal equinox to a star, LST

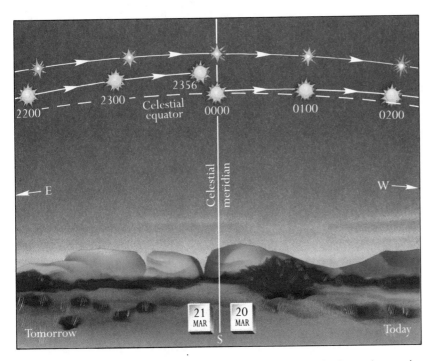

At noon (0000 hours), as it passes the vernal equinox on March 20, the Sun is shown with a star just above it. As the Sun and the star move west on their daily paths, the Sun moves north along the ecliptic to the east of the star. By the next day the Sun's declination is almost 1°N, and it has moved nearly a degree east of the star. The star therefore returns to the meridian not quite 4 minutes before the Sun, rendering the sidereal day nearly 4 minutes shorter than the solar day. (The movements shown are considerably exaggerated.)

must always be equal to the star's right ascension plus its hour angle. Obviously, then, $HA_{star} = LST - \alpha_{star}$, a relationship that serves to locate the star's hour angle. Follow the hour circle to the star's declination, and there it is. The system allows us to find our way in the sky and travel among the stars.

POLAR WOBBLES

Look at any modern star map or globe that shows the ecliptic. It will not have the vernal equinox in Aries but in Pisces, and will show the autumnal in Virgo. The twins, Gemini, firmly grip the summer solstice, and Sagittar-

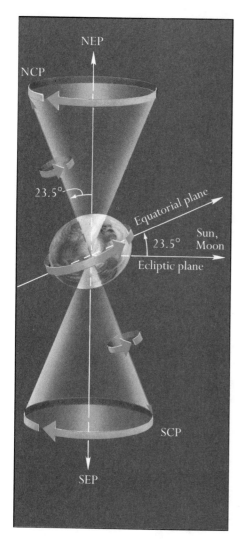

Since the Moon and Sun are not in the equatorial plane, their gravity exerts a torque on the Earth that causes the axis to precess. The wobbling axis takes the equator with it, causing the intersection of the celestial equator and ecliptic to regress.

ius balances the winter above his bow. What is going on? Why do we not speak of the Tropics of Gemini and Sagittarius? The anomaly is explained by a rotation of the entire coordinate frame relative to the stars produced by a wobble of the Earth's axis called precession, which takes 25,800 years to make a full turn. The Earth's axis simply rotates about the perpendicular to its own orbit, approximately maintaining the 23.5° angle but changing the positions of the celestial poles and equator in space. The physical cause is the gravitational pull of the Moon and Sun on the Earth's small equatorial bulge (about 20 km high and produced by terrestrial rotation). The result is the movement of the equinoxes and the solstices westward through the stars at a rate of about 50 seconds of arc per year, or about one zodiacal constellation every two millennia.

This knowledge is hardly new. It was discovered with the naked eye over 2000 years ago! In the third century B.C., the Alexandrian Greeks Timocharis and Aristyllus completed what was probably the first star catalogue to contain real measurements of angular positions. When Hipparchus compiled his catalogue a century and a half later, he found that the angles between stars and the equinoxes had changed. In the time of Hesiod and Homer, the vernal equinox *was* in Aries, whose symbol, ♈, is used to represent that point; indeed, the vernal equinox is still called the first point of Aries even though it is now in Pisces.

The reality of precession is wonderfully clear from a newspaper astrology column. The astrological signs of the Zodiac are tied to the vernal equinox, and when they were invented that point was in Aries, which usually starts off the list. But the sign of Aries now overlies the constellation Pisces, the fishes overlie Aquarius (which somehow does not change Aries' or Pisces' astrological attributes), and so on. The Sun may have been in Sagittarius when you were born, but if your birthday is between December 21 and January 19 you are still a "Capricorn"—whatever that means!

There are other fascinating effects of precession. Right now, the NCP points very close to Polaris, but it was not always so. Old Egyptian records refer to Thuban (γ Draconis) as the pole star. The pole will continue to get closer to Polaris until early in the 21st century and will then pull away; 14,000 years from now, Vega, which today passes overhead at New York, will be an acceptable polar marker. There is currently no pole star in the southern hemisphere. However, 3000 years ago, the Small Magellanic Cloud, a companion galaxy to ours visible to the naked eye, graced that spot.

As Vega appears to move toward the pole, stars that are now not seen from New York or Chicago—those that transit the meridian south of Vega but below the horizon—must begin to make their appearances. Alternatively, some stars that are now visible in those cities will appear to move toward the SCP, and if their declinations are already low enough will slip below the southern horizon. Native Americans saw the Southern Cross from North Dakota 6000 years ago; and if there are still Kansans in the year 12,000, they will see it again. Residents of southern Australia can now not see much of the Big Dipper, but let them wait 6000 years, and they will watch it pop up nicely above the northern horizon every winter.

If stars alter their angular distances from the poles, they must also change their declinations; and if the vernal equinox moves westward, the right ascensions have to change as well. Just when we thought we had arrived at a fixed set of coordinates, reality looms: Nothing is fixed, everything changes. The variation in coordinates is highly significant: 50 seconds of arc a year is huge when you consider that we can attain positional accuracies of 0.01 second! With appropriate equipment, the astronomer can see day-to-day variations even in a motion with a 26,000-year period. Still, the problem is not so great as it seems. From observation of the stars over long periods we can learn the equations of precession with high accuracy. We need only specify the right ascensions and declinations of stars for a specific moment, or epoch, and we can then perform the coordinate rotation to calculate the α and δ for the star tonight. Stars are commonly catalogued for the beginning of the year 1950.0, and with the new millennium, 2000.0 coordinates have become the standard.

At this point, precession appears to be quite simple, so perhaps as expected, the matter is not yet closed. The lunar and terrestrial orbits are not circular and the gravitational forces on the bulge are not constant. Furthermore, the Moon's path is inclined by 5° to the plane of the ecliptic so that the Moon and the Sun do not act in concert. These orbital attributes produce variations in the rate of precession that collectively are called the nutation. The chief terms are small oscillations with monthly and yearly periods, and a larger one (up to 19 seconds of arc) that results from an 18.6-year wobble of the Moon's orbital axis. There are over 100 known terms to the nutation, every one of which must be taken into account in order to locate the stars. Even more remarkable, we know what they are and we can correct for them.

TO KNOW THE STARS

Thoughtful examination of the heavens to make sense of their workings goes back almost to the dawn of civilization. Pythagoras dreamed about the sphericity of the Earth, Aristotle denied rotation, and elaborate mechanisms

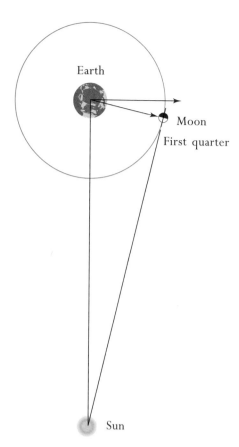

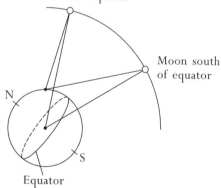

Aristarchus measured the ratio of the distances of the Sun and Moon (top) from the angle at which the Moon appears half illuminated (first quarter phase). Ptolemy found the lunar distance relative to the Earth's size by parallax (bottom), the difference in displacement seen when the Moon is north and south of the equator.

Earth

Moon
First quarter

Sun

Moon north of equator

Moon south of equator

N

S

Equator

were constructed to explain the movements of the planets through the skies and still have them going about the Earth. The issue is therefore not "rightness" or "wrongness" but the attempt to learn, out of which eventually springs the truth.

The key to understanding, however, is not pure thought, but examination and measurement. Look at what the early Greek astronomers were able to accomplish, consider their still-living legacy. Nearly 2400 years ago, Eratosthenes measured the size of the Earth, and only a century later Hipparchus discovered precession of the equinoxes. From here, we move out into space. Aristarchus of Samos, surely one of the more remarkable individuals of his time, devised an ingenious way of measuring the distance of the Sun relative to that of the Moon. If the Sun were an infinite distance away, the "elongation" of the first-quarter Moon, that is, its angle relative to the Sun, would be exactly 90°. The closer the Sun, the smaller the elongation. Aristarchus tried to determine it and obtained an angle of 87°, from which he found a distance ratio of 20. The true angle of 89°51′, which results from an actual ratio of 400, is impossible to measure because of the roughness of the lunar surface and the irregularity of the line on the Moon that separates day from night. Even though his value, in use until the seventeenth century, was wrong by a factor of 20, Aristarchus' concept was correct, and he clearly demonstrated that the bodies in the sky were not just set upon a sphere, but had a three-dimensional quality. Moreover, the Sun was indeed far away.

A century later, Hipparchus devised methods to estimate the distance to the Moon relative to the size of the Earth from the phenomena of solar and lunar eclipses. Some 300 years after that, the Alexandrian Claudius Ptolemaeus (Ptolemy) made a more direct determination from its parallax by observing lunar transits across the meridian at different points in the lunar orbit. Both men found that the ratio of the distance of the Moon to the Earth was about 30 terrestrial diameters, right on the modern value. These measurements, together with a value for the size of the Earth, allow the actual distance of the Moon and Sun to be found.

What of the stars? Aristarchus firmly held that they were much more distant than the Sun, but could not prove it. It was not until Copernicus demonstrated that the Earth (and other planets) went about the Sun, and that the geocentric Ptolemaic theory was wrong, that astronomers were in any kind of position to attack the problem: The observed lack of parallax now could be used to infer that the stars are truly at great distances. But with naked-eye observations, astronomers had gone about as far as possible. The watershed of astronomy, the great divide, was produced by a striking individual, Galileo Galilei. His first use of the telescope provided proofs of

Galileo Galilei (1564–1642) did not invent the telescope. He did something far more important: He learned how to use it.

the Copernican theory. Observation of lunar craters and mountains showed that other bodies in the solar system could be very much like the Earth, and the discovery of Jupiter's tiny orbiting moons demonstrated a clear analogy to the orbiting planets. The sight of Venus' phases put the strictly geocentric system to death: They could be caused only by our viewing different portions of daylight as the Earth and Venus swing about the Sun, and are not possible if everything goes about the Earth.

Moreover, Galileo had found the way to probe deeply into space, discovering that the mysterious Milky Way was simply made of more stars. His work, and that of his equally brilliant successors, would finally lead to the solution of the problem of stellar distances and demonstrate that stars are really distant suns like the one that lights our day. He provided the foundation that ultimately allowed us to learn what the stars really are, where they came from, what time holds for them, and since we are entirely dependent upon one particular star, what time holds for *us*. In order to follow these great advances, however, it is first necessary to explore Galileo's legacy, the telescopes and the instrumentation with which the discoveries have been made.

The 10-meter Keck telescope on Mauna Kea, Hawaii, is revealed as the shutter draws back at sunset.

Chapter Two

THE TOOLS OF DISCOVERY

HOW WE OBSERVE THE STARS

Early in the seventeenth century, several opticians were said to have noticed that combinations of lenses could make distant things look larger. The telescope was not Galileo's invention, but in his hands it took life. His real contribution was not so much that he looked and commented, but that he recorded and thought about what he saw. What followed was an explosion of discovery. Telescopes were made ever larger and more powerful. By the nineteenth century, astronomers were able to analyze starlight and thus could begin to dissect these distant suns. In our own century we have expanded the celestial science beyond light into radio, even X-ray, and have left the confining bounds of the planet to float our telescopes freely in space. This is the story.

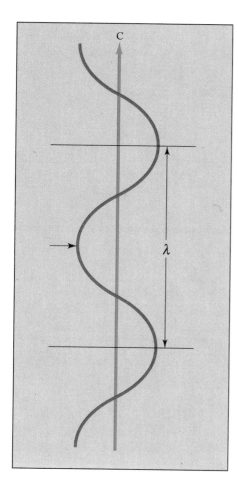

The wave is moving at the speed of light, c (300,000 km/sec). The wavelength is the distance between crests; the frequency is the number of crests that pass a particular point (the arrow) per second.

LIGHT

We cannot go to the stars; we can only observe and analyze, with the aid of telescopes, the radiation they send us. However, before we can understand how they operate we must examine the nature of this radiation and how it can be manipulated.

Light is generally thought of as an oscillating electromagnetic (EM) wave, similar in a loose sense to a wave moving in water. The difference is that light waves do not travel in a substance. In a vacuum, the waves move at a velocity, c, equal to 300,000 km/sec. Have respect for this number: A light beam could travel eight times around the Earth in 1 second and fly from here to the Moon in 1.3. It is the maximum speed allowable anywhere, a fact fully backed up by countless experiments and solid theory.

Any wave is characterized by two dependent numbers, wavelength (λ), the physical distance between crests, and frequency (ν), the number of crests that pass a point per second (measured in Hertz, abbreviated Hz). These two numbers multiplied give the speed, so in the case of light $\lambda\nu = c$. Light waves to which the eye is sensitive come in a variety—a spectrum— of wavelengths, which we identify as color. Red light has a wavelength of about 7×10^{-5} cm, or 7000 Ångstroms (1 Å$= 10^{-8}$ cm). As the waves become shorter, we progress from red through orange, yellow, green, blue, and then violet at about 4000 Å.

Nature does not stop at these limits. The entire electromagnetic spectrum stretches to over a factor of 10^{15} in wavelength, and different domains carry different names. The visual or optical region contains the observed colors. Longer than red lie the infrared and radio, in which wavelengths extend to kilometers. Shorter than violet, we encounter the ultraviolet, X-rays, and then gamma (γ) rays, the last with wavelengths less than an Ångstrom. In a vacuum, all of them move at c. The stars and their associated phenomena send radiation across the entire spectrum, and to understand processes in space, we have to be able to observe it all.

The wave description of light (commonly used as a synonym for all EM radiation) is strangely inadequate. In some circumstances, light behaves not as a wave but as a particle. Such a particle, which also has the wave properties described above, is called a photon. Light is the principal way by which energy (the ability of a body, or radiation, to do work on another body, to accelerate it over a distance, or to heat it) is carried through the Universe. The energy transported by a particular photon depends on its frequency according to the equation $E = h\nu$, where h is a universal constant of nature called Planck's constant. Gamma-ray photons then carry some 10^{15} times as much energy as radio photons, which is why a radio

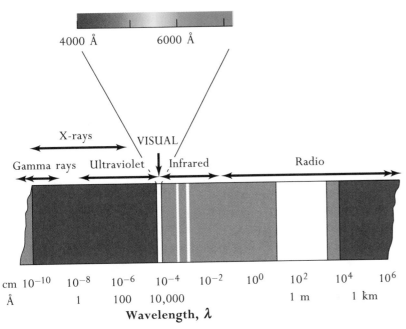

4000 Å   6000 Å

X-rays

VISUAL

Gamma rays   Ultraviolet   Infrared   Radio

cm 10^{-10} 10^{-8} 10^{-6} 10^{-4} 10^{-2} 10^{0} 10^{2} 10^{4} 10^{6}

Å   1   100   10,000   1 m   1 km

Wavelength, λ

The EM spectrum, which runs from wavelengths shorter than 10^{-8} cm to kilometers, is loosely divided into broad, overlapping bands. Since many physical processes do not radiate energy in the optical, the only way to know of them is to study the relevant spectral regions. The lighter the shading, the greater the transparency of the Earth's atmosphere.

broadcast will not hurt you but an atomic bomb's γ rays will. The amount of energy transported by a single photon is quite small. For yellow light, $\lambda = 5500$ Å, so $\nu = 5 \times 10^{14}$. Since h (in common meter-kilogram-second units) is a mere 6×10^{-34}, E is only 3×10^{-19} joules (a joule is the unit of energy). A hundred-watt light bulb radiates 100 joules per second, so (very crudely) it sprays an amazing 10^{20} photons at you every second.

OPTICAL PRINCIPLES

Light interacts with matter in all sorts of ways. For example, it can be reflected. A ray of light aimed at a smooth reflecting surface, a common mirror, will bounce off at the same angle (measured against the perpendicular) at which it arrived. If not for that simple principle, your face would appear weirdly distorted in a bathroom mirror.

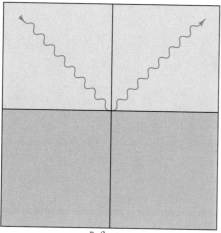

Reflection.

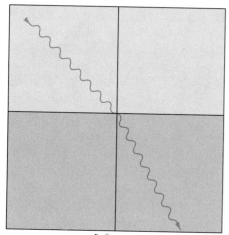

Refraction.

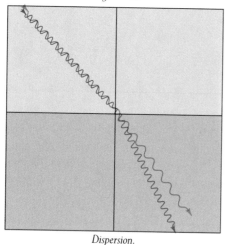

Dispersion.

A light ray can also be bent by passing it through a substance. Send a beam hurtling at an angle from the air to a transparent material such as glass or water. As it enters, it slows, the change in velocity causing it to bend, or refract, toward the perpendicular. The greater the angle of incidence, the more the refraction. If the ray goes in the other direction and *leaves* the water, it will bend in the opposite direction—*away* from the perpendicular—as it speeds up. The result is that anything under water will appear misplaced and distorted (just look at the fish in a tank). The degree to which the light is slowed and the beam refracted depends on the substance. The speed of light, c, divided by the internal velocity is called the index of refraction, n. Glass, with n equal to about 1.5, has a higher index than water, and diamond has the highest of all, nearly 2.5.

The index of refraction depends on wavelength and almost always increases as λ decreases. The result is that violet light is refracted more than green, and green more than red. When a ray of white light, which is a mixture of all the colors of the spectrum, enters or leaves a substance, rays with different wavelengths will bend in slightly different directions, and the white light becomes dispersed into its components. Dispersion is everywhere around you—in a rainbow, in colored sparkles from crisp snow, in the shower of hues from a well-cut diamond. It is the phenomenon that allows us to take starlight apart and learn how stars are put together. Pass a sunbeam through a prism, a triangular piece of glass, and it is dispersed both upon entry and exit to show the colors of the Sun.

Light can also be bent and dispersed by a more obscure phenomenon called diffraction. Send a beam of monochromatic (single-color) light at a plate with a pair of slits cut into it. Each slit acts as a new source, and the waves leave them in step—that is, in phase. When the light from the apertures falls upon a distant screen, the different sets of waves interfere with one another, producing a set of light and dark bands called fringes. The brightest fringe is centered exactly on a line between the two slits at a point where the distance from the fringe to each of the slits is the same. The two sets of waves then overlap one another and add together, a condition called constructive interference. A small distance away on either side of the first fringe the distance from one slit is a half-wavelength greater than the distance to the other. The crest of one wave then fills in the trough of the other, producing destructive interference and a zone of darkness. Farther away yet, the difference in distance is a full wavelength, and the waves again overlap and constructively interfere. The pattern continues through 3/2 wavelengths, 2 wavelengths, and so on. Remarkably, interference fringes are also seen with a single slit. In a sense, the single slit is made of an infinite number of pairs of double slits, all so close together that they

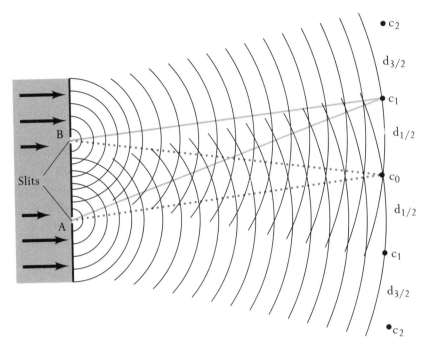

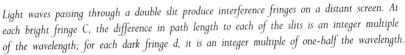

c_2

$d_{3/2}$

c_1

$d_{1/2}$

B

Slits

c_0

$d_{1/2}$

A

c_1

$d_{3/2}$

c_2

Grating spectra.

Light waves passing through a double slit produce interference fringes on a distant screen. At each bright fringe C, the difference in path length to each of the slits is an integer multiple of the wavelength; for each dark fringe d, it is an integer multiple of one-half the wavelength.

produce a single interference pattern. You can see these fringes if you place your fingers very close together next to your eye and look at a bright light.

Diffraction also produces spectra. The greater the wavelength, the farther from the center a bright fringe will be. If white light falls upon the double slit, each of the fringes will have red light falling on one edge and blue or violet on the other. If a third slit is added, the conditions for constructive interference are more stringent, and the color separation will be clearer. The more slits, the better the spectra. If thousands of slits are used, the spectra are stunning. A device that produces such spectra is called a diffraction grating. Usually these are reflective plates with grooves cut into them. The most common example is a compact audio disk, which has a fine spiral groove. Hold one up to the light to see several spectra, one after the other. Spectra made this way are superior to those produced by prisms because diffraction spreads the light linearly relative to wavelength, while prisms compact redder light. Moreover, glass absorbs in the ultraviolet.

Finally, the wavelength and frequency of EM radiation depend on the relative velocity along the line of sight between the source and the observer, a phenomenon called the Doppler effect. If you move into a set of waves, such as those on a lake, the frequency with which you encounter them will be increased and their separations will appear to be diminished; if you move along with them you experience the opposite. The effect is common with sound. An airplane approaches at high velocity, and the pitch of the engine is higher than it would be at rest. As it passes overhead, the line-of-sight speed drops, and so does the pitch; as the craft recedes, the wavelengths become longer and the pitch drops even further. If a light source is coming toward you its radiation will be shifted to shorter wavelengths; that is, it will appear to be slightly bluer than it would be if the source were at rest: Recession will produce a shift toward the red end of the spectrum. The relative amount of the shift is directly proportional to speed, or $\Delta\lambda/\lambda = v/c$ (the Doppler formula), where λ is wavelength, $\Delta\lambda$ is the change in wavelength, and v is the line-of-sight speed between the source and observer.

The formula works only if v is much less than c; if v approaches c, a different equation based on the theory of relativity must be used. In most instances the speeds and the shifts are very small. The color of a source can change only if the source is moving at a significant fraction of c. A common example of the EM Doppler effect is police radar. A radar gun transmits a radio signal at a precisely defined wavelength. When the radio waves bounce off an approaching car, their wavelengths are made shorter. The radar gun picks up the returned signal, compares its wavelength with that sent out, applies the Doppler formula, and (perhaps dismayingly) displays the vehicle's speed.

With these principles in hand we can build the tools necessary to study the stars.

REFRACTORS

Even in its most sophisticated form, a telescope is extremely simple in its basic operation. It is a collector of light with two distinctly different purposes: to make images of faint sources bright so that there is enough light for analysis, and to provide resolution, which is the ability to see the separation between things that are close together. Galileo's telescopes used lenses and employed the principle of refraction. Stars are effectively at infinity, and the light waves that fall upon the Earth are parallel to one another. They are bent by refraction at the spherically curved surfaces of the lens toward a focal point, where they meet. If we place a detector such as a

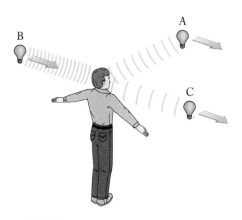

Light bulb A is moving past the observer, but because there is no motion along the line of sight, the observer sees the original wavelength. Bulb B is coming at the observer, who sees the waves Doppler-shifted to shorter wavelengths. Bulb C is moving away, and the waves are shifted longward. The effect is the same whether the source or the observer is moving.

photographic plate in the plane parallel to the lens that holds the focal point, we can record an image of the star. That is all there is to it: a lens and a detector.

The most critical parameter of a telescope is the diameter, or aperture, of the "objective," the light-gathering lens. The larger the aperture, the more light it can accumulate, and the brighter will be the image and the fainter the stars that can be seen or photographed. The amount of light that can be collected is proportional to the area of the objective lens, and consequently to the square of the aperture. Aperture is everything. Telescopes are named by aperture: We speak of the "Palomar 200-inch," (or "5-meter"), the "Kitt Peak 4-meter," or the "Lick 36-inch." The dark-adapted human eye has an aperture of about 7 mm. With a 70-mm telescope you could view stars of the 11th magnitude—100 times fainter than those you can see without aid.

Of course, with the telescope we see and record not one star but a whole field of them over a specific angular diameter. The scale of the field of view depends not on the aperture, but directly on the focal length (the distance from the lens to the focal point). The longer the focal length, the more widely separated two given stars will be. The single lens will also invert the image on the focal plane, east for west, north for south.

Galileo did not have photographic film. He looked through the instrument, just as the average backyard astronomer does today. If you were to place your eye in the focal plane, however, you would not see stellar images but instead the illuminated objective. In order to see the stars, you must use an eyepiece, which renders the rays parallel once again so they can be focused on the retina. There are two choices. Galileo placed a concave lens in front of the focal plane. This method results in a normal, uninverted image and is still used in inexpensive binoculars and opera glasses, but the resulting magnification and field of view are quite restricted. It is far better to use a convex lens behind the focal plane. In a loose sense, the eyepiece is used as a magnifier to examine the focal plane. The resulting angular magnification, or power, of the combination depends on the ratio of the focal lengths of the objective and the eyepiece, or $M = F_{objective}/F_{eyepiece}$. The power is easily changed by switching eyepieces. The one disadvantage of this system is that it still inverts the image. Astronomers get used to looking at things rotated upside down.

The professional astronomer rarely looks through the telescope. Direct viewing is done only for target acquisition and for guiding during long exposures, and even that can be accomplished automatically. The instrument is used instead to feed light to some kind of detector. Therefore, magnifying power is of little importance. What is significant is scale, and

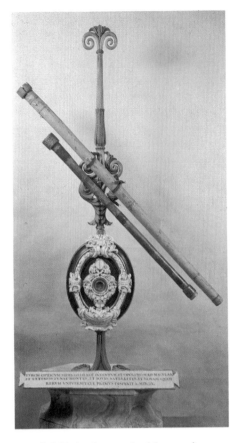

Two of Galileo's telescopes; he did extraordinary things with instruments that were of lesser quality than the cheapest you will find in the average discount store.

The earliest telescopes were refractors. The curved surfaces of the lens (left) bend the incoming parallel stellar light rays and bring them to a focus at the center of the focal plane, where an image of the star is created. A photographic plate placed there will record a picture of the star. The brightness of the image depends on the square of the diameter, or aperture, of the lens. The light from two stars enters the lens (right) and is brought to a focus at the focal plane. The scene is inverted on the focal plane: up for down, left for right. An eyepiece placed behind the focal plane renders the light rays parallel once again and magnifies the angular separation of the stars. The field of view is still inverted.

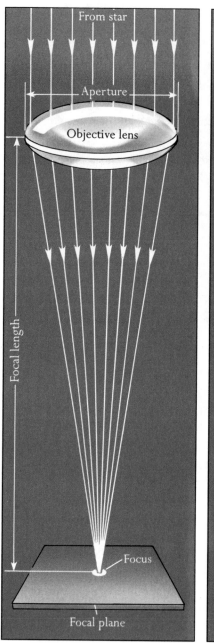

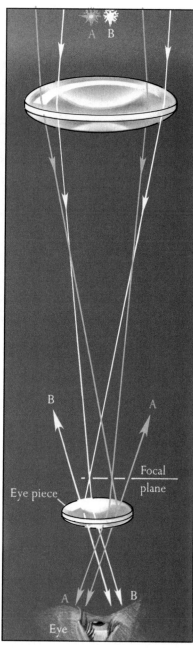

more important, resolution (also called resolving power), which involves diffraction. A telescope aperture acts as a single slit, and although it is very wide, it still produces an interference pattern. The stellar image is the bright central fringe. Since the lens is circular, this image is surrounded by a series of light and dark circular fringes. Although the star is effectively a mathematical point so far as the telescope is concerned, the central fringe, or diffraction disk, is not. The ability of a telescope to resolve fine detail depends only upon the diameter of this disk, which shrinks as aperture is increased. The only way to achieve a point image from a point source, however, is to make the aperture infinitely large, which is impossible. As a standard rule of thumb, the resolving power of a telescope in seconds of arc is $5''/D$, where D is in inches, or $13''/D$ if D is in centimeters.

Practically, however, resolution is not established by the telescope but by the Earth's atmosphere. The charming phrase "twinkle, twinkle, little star" is enough to spoil an astronomer's whole day, let alone night. The twinkling, or "seeing" as it is properly called, is caused by variable refraction in the air, and it smears the apparent point that is a star into a disk that can be several seconds of arc across. Twinkling is one of the principal reasons why professional telescopes are placed on high mountains, where the seeing is the best. But even at the choicest sites, the image size, or "seeing disk," never gets below about a half-second of arc. Once the aperture is above about 12 inches (30 cm), gains in resolution become increasingly limited without using highly advanced equipment designed to overcome the effects of atmospheric seeing.

Refractors have three debilitating problems. First, refractive dispersion causes the blue and violet rays from a star to converge faster to a focus than the red rays. Consequently, the location of the focal plane depends on color. If you try to focus on the red or yellow image of a star, there will be a big blue haze around it. The effect of this chromatic aberration is minimized by long focal length, and even better controlled by constructing the objective out of two or more elements. An achromatic objective (something of a misnomer) usually consists of two lenses with different curvatures and different indices of refraction placed together. In that way any two colors of choice can be brought to a point. Three colors can be brought to a focus by three lenses, but each added lens also absorbs and reflects some light, so except for some specialized cameras, two is generally the limit. Residual chromatic aberration is reduced by using a large focal length and filters that block the unfocused light.

The second problem is aperture. A refractor objective uses two lenses of high optical quality and has four surfaces to be ground and polished, and the cost goes up dramatically with size. The largest refractor ever built is only about 1 meter in diameter. Moreover, the tubes have to be so long that the supporting structure and the observatory dome must be huge and massive, adding further to the already great cost. The third problem is that

The images of stars made by telescopes can be quite complex. The central image of this bright star is actually a diffraction disk surrounded by circular fringes (smeared out in this photograph by atmospheric effects). The rings are produced by reflections within the telescope system. The crossed spikes on the image are true diffraction effects caused by the struts that support a secondary mirror.

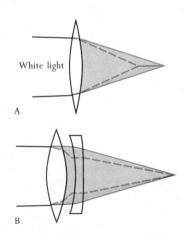

The image of a star made by a single lens will be a colored line along the optical axis, because blue rays bend more than red. Only a single color can be brought into focus at a time. A combination of convex and concave lenses, with different indices of refraction, can bring any two colors to a focus.

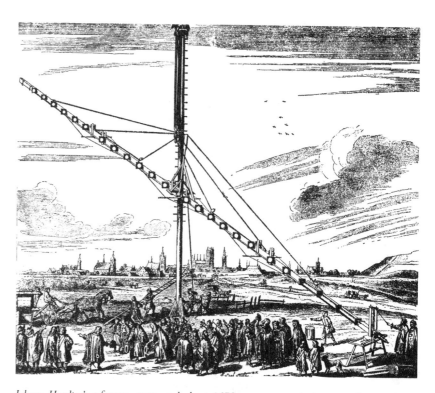

Johann Hevelius's refractor, constructed about 1670, was enormously long in order to reduce chromatic aberration. It is hard to see how he found anything in the sky with it.

The 24-inch refracting telescope of the Lowell Observatory points, through the slit in its dome, at Mars.

glass absorbs the ultraviolet radiation that gets through our atmosphere, seriously reducing the instrument's research capability. The problem can be ameliorated with special—and costly—ultraviolet-transmitting glasses, or by using quartz, which is more expensive yet. Yet for all the difficulties, it is a testimony to the beauty and simplicity of these magnificent instruments that although the last great refractor was built nearly a century ago, they can still be used effectively for specific types of research.

THE REFLECTOR REVOLUTION

The solution to these difficulties was initiated in the seventeenth century by none other than Isaac Newton (1642–1727), who invented the reflecting telescope, which uses a mirror instead of a lens to focus light. It was not until the mid-1700s, however, that the technology was sufficiently developed that the reflector could compete with the refractor. Then, in the hands of William Herschel (1738–1822), who effectively founded modern astronomy, the reflector came alive. In this century, it finally won out.

Since the angle of reflection does not depend on color, there is no chromatic aberration, and since there is no transmission of light through anything, there is no ultraviolet absorption. Because there is only one optical surface to be ground, the interior of the mirror can even be flawed. The instrument can be made very large and still quite short, reducing the size of the containing building. As a result, reflectors are much cheaper to build than refractors, freeing money that can be used to increase the size.

A telescope objective mirror begins with a concave circular disk of glass or ceramic, the latter now used to reduce thermal expansion effects. A spherically shaped surface will not properly focus light, however, because radiation from the outer zones of the mirror comes to a focus before that from the inner zones. The resulting spherical aberration produces a fuzzy image. Instead, the mirror is worked into the shape of a paraboloid, which will bring all parallel light rays of a star on the optical axis to a point focus. The surface facing the incoming light is overcoated in a vacuum chamber with a very thin, bright layer of aluminum. The focal plane, or the prime focus, is in front of the mirror. How does the observer avoid blocking the light? Newton's solution was to place a small, flat secondary mirror at a 45° angle just in front of the focus, sending the light to the side. This Newtonian focus, common in older reflectors, is popular for amateur telescopes; it is inexpensive and places the viewer in a comfortable position.

There are a number of other possibilities. Most professional telescopes use the Cassegrain focus, in which light is reflected from the secondary through a hole drilled in the objective, or primary, mirror. The secondary is ground into the shape of a convex hyperboloid, making the light converge at a smaller angle and sending it to a focus that, like the Newtonian, is outside the tube. By adjusting the curvature and placement of the secondary, very long *effective* focal lengths—far greater than that of the primary—can be achieved in a very short tube. Moreover, the secondary is easy to change, so different focal lengths (and consequently different focal scales) are available for different purposes. The objective may have a focal ratio

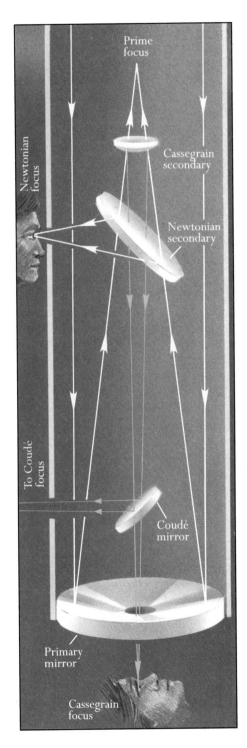

The reflecting telescope uses a large, paraboloidal primary mirror to collect and focus light. The prime focus is often used for direct imaging because nothing intervenes in the light path. Amateur instruments commonly employ the Newtonian focus. Large equipment—a spectrograph, for example—requires a more mechanically stable platform, so the Cassegrain focus is often used. For massive equipment, the light is further reflected to the coudé focus.

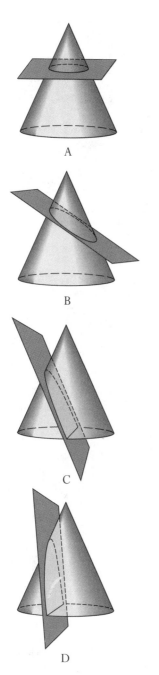

The conic sections are a set of curves produced by cutting a cone with a plane: (A) circle; (B) ellipse; (C) parabola; (D) hyperbola. Rotation of the parabola around its axis generates a solid paraboloid, the shape of a primary telescope mirror.

(focal length divided by aperture) as short as 3 (called f/3) and effective Cassegrain focal ratios of f/7, f/13, and f/30 or greater. Moreover, the focal position is on the mechanical axis of the instrument, allowing easy placement of heavy analyzing equipment.

In another variation on the theme, the light from the Cassegrain secondary can be intercepted by a flat mirror placed in front of the primary at one of the rotation points of the instrument. It is then shot to the side where additional mirrors send it to a fixed room in the basement of the observatory to the coudé focus (from the French for "bent"), where massive equipment can be mounted. Finally, if the telescope is large enough, we can actually use the prime focus. Several giant reflectors have a cage built into the instrument at that position, allowing the observer to ride inside them during the night. All these options make the reflecting telescope enormously flexible while remaining relatively inexpensive.

The first of the really great reflectors, completed in 1918, was the 100-inch telescope built by the Carnegie Institution of Washington on Mount Wilson above Los Angeles. It was followed 30 years later by the Hale 200-inch (5-meter), constructed by Carnegie and the California Institute of Technology on Palomar Mountain about halfway between Los Angeles and San Diego. Just think of the light-gathering power of this telescope.

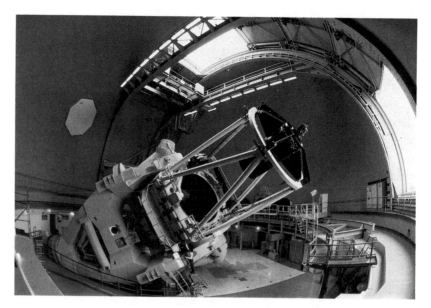

The 3.6-meter telescope of the European Southern Observatory (ESO), in La Silla, Chile.

The 10-meter Keck telescope under construction atop 4200-meter (13,800-foot) Mauna Kea.

It will collect over half a million times as much light as the unaided eye, and a view through an eyepiece would allow you to see 20th magnitude! A marvel of engineering, it is so delicately balanced on high-pressure oil pads that all its 500 tons can be moved by hand. Since its construction, several telescopes in the 3- to 4-meter class have been built, but except for a largely ineffective 6-meter in the Soviet Union, only recently has the Hale 200-inch been surpassed in size. A number of 8-meter instruments, with 2.5 times the light-gathering power of Palomar—a full magnitude of brightness—are planned. It is unlikely that the world will see monolithic mirrors much larger because of the difficulty of casting, polishing, and mounting them. The huge Keck 10-meter telescopes now operating in Hawaii consist of segmented mirrors, each individual reflecting surface only a meter across. With such innovative designs, there is no immediate limit to how large telescopes might become except that imposed by available money (a *very* significant constraint).

Reflectors have their own problems. A simple paraboloid will produce a beautiful image, but only on the optical axis. Stars that are not right in the center of the field of view suffer from an aberration called coma (from the Latin for "hair"), which draws the stellar images into diffuse lines that point radially outward. The effect greatly restricts the field of view: It would take lifetimes to survey the entire sky with the Palomar 5-meter. One solution was devised by Bernard Schmidt in the 1930s. The instrument he developed uses a nonfocusing objective lens called a corrector plate that is placed in front of a larger spherical mirror. The complex shape of the corrector plate changes the paths of the light rays to compensate for both spherical aberration and coma. The result is a camera with a huge field of view. The most famous of these instruments is the 48-inch Schmidt at Palomar. Another solution, the Ritchey–Chrétien design (used in the Keck), employs hyperboloidal primary and secondary mirrors.

DESIGN AND OPERATION

All telescopes must rotate on two axes in order to point anywhere on the two-dimensional surface of the celestial sphere. In the standard equatorial configuration, one axis points to the celestial pole, permitting the telescope to rotate along a daily path parallel to the equator. The other axis allows it to be pointed north and south, perpendicular to the equator. Each axis has a setting circle, one reading hour angle, the other declination. In classical or historic instruments these are iron rings with engraved graduations. The astronomer looks up the coordinates of a star, planet, or galaxy. The difference between α and the LST from the observatory clock gives the hour angle. When the positions are set, the object appears in the eyepiece of a low-power finding telescope attached to the telescope tube. When the observer centers the object in the finder, it appears in the main instrument. The polar axis is directly attached to a sidereal clock, which slowly moves the telescope along a parallel of declination, keeping the star in the field of view as the Earth rotates.

The telescope is generally housed in a rotating dome with a shutter that can be opened to the sky. The dome provides protection against inclement weather and, when the shutter is open, against wind and stray light. Telescopes are outdoor instruments that must be closely maintained at the ambient temperature in order to minimize the size of the seeing disk. No heat is allowed, and in the past the astronomer often endured rigorous working conditions. Temperature control is so critical that the outside of the dome is painted with a special titanium dioxide compound that reflects sunlight and radiates well in the infrared. Some telescopes even have chillers built into the floor so that daytime temperatures can be held low.

Professional astronomers have largely removed themselves from the telescope. The eye was long ago replaced by dispassionate instruments, except perhaps for an irresistible peek at a favorite object during twilight (the view through a 3- or 4-meter telescope is awesome). Setting circles have also been succeeded by electronic encoders that give the precise rotation of the telescope shafts. Just type α, δ, and the epoch onto a keyboard. A computer will perform the precessional rotation to determine tonight's coordinates, read the LST from its internal sidereal clock, do the subtraction, and command motors to turn the rotational shafts to the desired setting. The computer contains information that can compensate for the flexure of the heavy telescope and for the lofting effect of refraction produced by the Earth's atmosphere. The control is so precise that the finder is not needed: A modern instrument can be set, and will track or follow a star, to the amazing precision of 1 second of arc. The sought-for object then appears almost magically at the center of a television monitor, allowing the astronomer to direct the light to the analyzing instruments. Since all this is done automatically, the astronomer no longer need be in the dome at all and can sit comfortably in a "warm room" sipping hot coffee while the data are acquired.

Computer control has also freed us from equatorial design. Because the engineering is far simpler, it is less expensive to align a telescope mounting to the horizon so that one axis points to the zenith along the direction of the Earth's gravity. The computer continually converts the right ascension and declination to altitude and azimuth (the angle along the horizon measured from the north point through east), moving the telescope smoothly through the sky. All new large telescopes, such as the Keck, work this way.

Large telescopes are so costly and their design and control so complex that they must be run by highly trained operators. The astronomer provides the coordinates and is given command only of slow motions for guiding.

ANALYSIS

By itself, the telescope does little. Its purpose is to feed light into analytical detectors. Until nearly the end of the nineteenth century, all we had was the human eye, which cannot make permanent records. The first great breakthrough was photography. The astronomical camera is no more than a plate-holder. It is affixed to the focus, the shutter held open sometimes for hours to integrate and record a faint signal as the telescope tracks through the sky. Astronomical photographic emulsions are almost always spread on glass plates, which provide great durability and dimensional stability. Sophisticated chemistry—mostly by Eastman Kodak—has gone into producing emulsions that record well at very low light levels. The plates range in size from small chips a centimeter wide to monsters half a meter across used in Schmidt systems. Photography has been the astronomical workhorse of the twentieth century. In the 1950s, the Palomar Schmidt was used to survey the entire sky visible from southern California in two colors, blue and red, down to about the 18th magnitude. (Although some magnificent color photography is shown in this book for illustration, research photography is always done in black and white, the different "colors" referring to filters used to make the images.) A companion instrument run by the United Kingdom and the European Southern Observatory in Australia has now done the same in the southern hemisphere, and the Palomar Schmidt is redoing the northern survey in *three* colors. These are all mammoth undertakings, search missions hunting for objects of interest to be examined by the big reflectors.

The photographic plate has some severe disadvantages. It is very inefficient, recording at best only about 2 percent of the light that falls upon it, and enormously long exposure times are required. It is also nonlinear. The blackening rate of the photographic grains does not respond directly to the intensity of radiation, so careful calibration is required. The result is considerable inaccuracy in the measurement of brightnesses. Very gradually, photography has been replaced by electronics. The revolution, which is now nearly complete, began about 1915 at the University of Illinois with the work of Joel Stebbins and the invention of the photoelectric photometer. This device makes use of the particle nature of light. A variety of substances will release electrons when struck by photons (the particles of light that carry energy), with an efficiency near 90 percent. These electrons can be accelerated by an electric field to produce a measurable current that is directly proportional to luminosity. In Stebbins's first successful experiment he was able to record the Moon; 50 years later, photoelectric photometers could measure the brightness of a 20th magnitude star.

The photoelectric photometer can observe only one star at a time and, although accurate, is again inefficient. Nor does it speed the process of imaging the sky. Video cameras used in the 1970s were an advance, and the ultimate is the charge-coupled device, or CCD, which also has an efficiency near 90 percent. The surface of the CCD is divided into tiny picture elements. When exposed to light, each of these pixels will build up an electric charge that is directly proportional to brightness: that is, each photon of light will produce the same slight increase in charge. The charge pattern in the chip will thus match the brightness pattern of the field of view. When the exposure is complete, the charges are drained out as an electric current, a row of pixels at a time. The computer reads the fluctuating current at a precisely timed rate so that each instant can be related to a specific pixel. The pattern is stored on disk or tape and can be reassembled, row by row, column by column, to re-create the image on a video monitor. A modern CCD system can record in one minute what used to take one to two hours with photography, enormously increasing the efficiency of the telescope and allowing us to reach an astounding 29th magnitude with the largest instruments. Moreover, the electronic linear recording permits us to measure the magnitudes of all the stars in the field of view at once.

The electronic revolution has allowed small 1-meter telescopes to do what was impossible a mere 20 years ago even for the giant 5-meter. That is the principal reason why the Palomar instrument was not exceeded in size for so many years. We were getting effectively larger and larger telescopes without having to build them. Now that we have reached nearly full efficiency in detection, we have to put our efforts back into glass and iron.

Photography, although down, is not yet dead. CCD chips are small, a 2048-by-2048 array the largest yet constructed. A mere 30 mm on a side, they can still image only restricted fields of view. Wide-angle cameras like the Schmidts must still resort to plates. But CCDs have so taken over the astronomical industry that Palomar has now banned photography from the 5-meter, a momentous historical watershed.

All observatories are equipped with far more than imaging devices. The spectrograph, for example, disperses light rays (rendered parallel by a collimator) into their component colors or wavelengths to record spectral signatures of individual atoms and to measure velocities. Prisms will do the job, but all modern spectrographs use the much more versatile diffraction gratings. In older days, the spectrum was recorded on a photographic plate. Spectrographs now come with their own CCD systems that allow the recording of spectral information on magnetic disk or tape.

Observing is actually the simple part; more difficult is the reduction of the images, magnitudes, or spectra to meaningful data. CCD pixels are not of uniform sensitivity, so the brightness of one part of an image may be distorted relative to that of another. The Earth's atmosphere dims the incoming radiation, making stars look fainter than they are, the effect dependent on altitude. The air also absorbs better in the blue than in the red and consequently garbles the relative brightnesses of the different colors in the spectrum, as do wavelength variations in the total transmission efficiency of the telescope optics. All these effects must be corrected for by various calibration techniques. One night's worth of good data can require several days at the computer terminal. Then, of course, the real work begins—the interpretation of the data, which might take the next year or even decade.

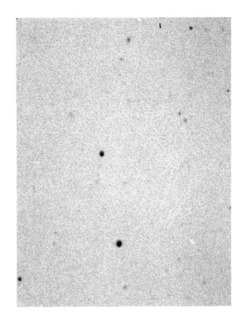

The sky photographed with an exposure time of 45 minutes.

A CCD image exposed for 125 minutes; it picks up images vastly fainter than the photograph would have done in the same length of time.

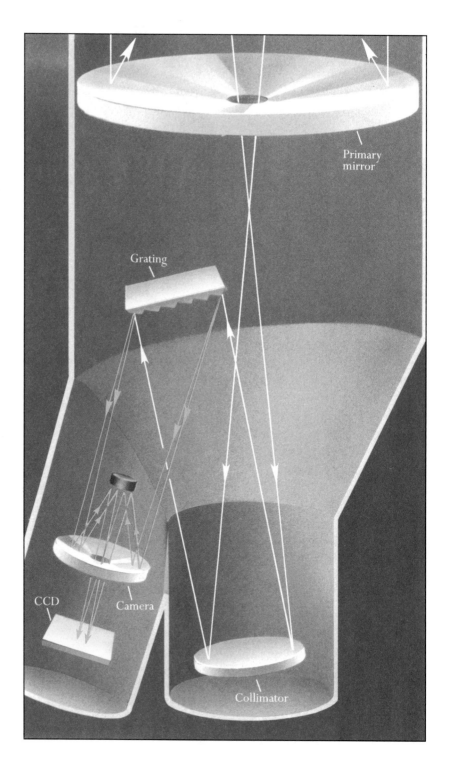

A spectrograph records stellar spectra. The light is first directed to a collimator that renders the rays parallel, then to a finely ruled grating for dispersal into the spectrum. A camera focuses the spectrum for imaging on the surface of a CCD.

Observatories work on a tightly scheduled and organized basis. There are many astronomers and few large telescopes, so it is necessary to apply for time as much as six months to a year in advance for a few precious nights. The analyzing equipment is usually so complex that it is mounted on the telescope for an extended period, days or even weeks at a time, with various astronomers rotating in and out to use it. The scheduling is done in close synchronization with the Moon, which can light the sky so brightly that faint objects are unobservable. Imaging of weak sources, for example, is generally reserved for the "dark run," the two-week interval centered around the new Moon, while bright objects are observed during the "bright run," the fortnight centered on the full Moon. In the dim ages when astronomy began, people were fascinated by the comings and goings of our satellite. Even the most modern of astronomers still pay homage to it.

RADIO TELESCOPES

The electromagnetic spectrum is enormously wide. Different physical processes radiate energy in different wavelength domains, and if we confine ourselves only to the narrow optical band of the spectrum, we severely restrict our view of the Universe. The first nonoptical spectral region was opened to the sky in 1933 by a Bell Laboratories engineer named Karl Jansky, who had been assigned to track down sources of radio interference. He found that the radio noise background swelled and fell with a sidereal period of $23^h\ 56^m$, a sure tipoff that at least one of the sources was extraterrestrial. It turned out to be the Milky Way. Jansky had unwittingly built the first radio telescope.

In its simplest form, a radio telescope is only a directional antenna—a long wire or a TV dipole is an example—held up to the sky. Radio waves from space excite an electrical current in the antenna, which is then amplified and measured. However, a bare antenna cannot gather much energy: Its light-gathering power is low and its directional capability poor. In order to collect a large amount of energy, the antenna is commonly placed at the focus of a paraboloidal reflector. The radio telescope is then the exact analogue of an optical instrument in which the detector, instead of being a photographic plate, a photomultiplier, or a CCD, is an antenna coupled to an amplifying system.

As in an optical instrument, this paraboloid produces a diffraction pattern. The size of any interference pattern depends directly on wavelength. The longer the wavelength, the farther away the first bright fringe

Inside the control room of the world's largest steerable radio telescope, located at the Max Planck Institute, Effelsburg, Germany.

and the larger the central disk. As a result, simple radio telescopes have notoriously poor resolution and in compensation must be made very large. One operating at a fairly typical wavelength of 10 cm would have to be nearly a mile across just to achieve the resolving power of the unaided human eye! Such a device is impossible to engineer and construct. The largest single steerable radio dish ever built is "only" 100 meters across, and operating it is something akin to balancing an athletic stadium on twin axes and moving it across the sky. The champion is a fixed 1000-foot-wide parabola built into a valley in Puerto Rico.

Fortunately, the problem of limited resolution is easily overcome. Imagine a mile-wide telescope. Cut out the entire middle and replace it with two separate telescopes that are a mile apart and mix or correlate their output signals. As a radiating source is tracked across the sky, the joint telescopes act like a double slit that "sees" a changing pattern of interference fringes. The interference pattern can be mathematically inverted to reconstruct the field of view. This simple "interferometer" necessarily will lose information, as it is constructed in only one dimension. But if we fill in the whole surface of our hypothetical mile-wide instrument with an array of individual telescopes, each of which is easy to build, we can accurately synthesize the effect of having observed with the big one. The largest of these, the Very Large Array, or VLA, is assembled on the plains of New

The Sun sets behind the radio telescope of the Parkes Observatory, New South Wales, Australia.

Mexico. It consists of 27 25-meter antennas mounted on railroad tracks and can synthesize a single telescope 36-km wide, the size of the Washington, D.C., beltway! With it, astronomers can achieve resolving powers that are actually better than those of optical instruments, since radio waves are largely unaffected by the Earth's disturbing blanket of air.

The VLA is the practical limit for electronically linked telescopes. The lengths of the electrical cables that tie each one to the central correlator cannot be accurately matched for interferometers much larger. To get around this difficulty we disconnect the telescopes and synchronize their observations with precisely matched atomic clocks. The data are merged and the interference fringes re-created in the computer. There is then no limit on size. The Very Long Baseline Array (VLBA) has receiving elements that stretch from New England to Hawaii, thus effectively creating a telescope that is 8000 km across, allowing a resolution of better than a hundred-thousandth of a second of arc! Even longer baselines have been established between Europe, the Soviet Union, and the United States, creating a telescope almost as big as the Earth. Someday, we will have an element of an interferometer array on the Moon.

A radio telescope can be used as an imaging device. The instrument can scan across the sky over a source of radiation and build up the spatial pattern of its emission. Then the appearance of the source can be recon-

The Very Large Array is spread over an area 41 kilometers across near Socorro, New Mexico.

structed in the computer so that the image looks just like a photograph, one seen with radio eyes. Radio telescopes are also outfitted with spectrographs and can be tuned much like a home radio—which is in fact a form of spectrograph—across the radio spectrum to measure changes in the intensity of the signal.

INTO SPACE

Despite the sophistication of modern telescopes, they are severely limited by their attachment to the Earth's surface. Our atmosphere transmits radiation only in limited portions of the electromagnetic spectrum. The infrared is heavily absorbed by carbon dioxide and water vapor, and there are only a few fairly narrow bands through which it is possible to look into space. The ultraviolet is taken out by ozone absorption. Only a small portion of the near ultraviolet, that just shortward of the optical, sneaks through. For life on Earth, these absorptions are critical. Absorption in the infrared acts as an insulating blanket that keeps us warm (the famed greenhouse effect) and

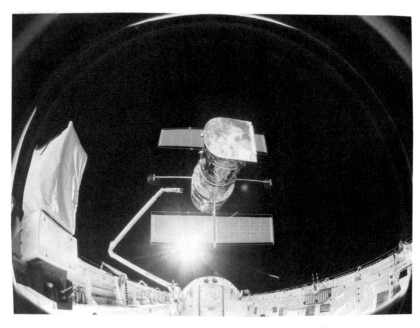

The Hubble Space Telescope, *still tied to the shuttle arm, hovers above the Earth. It now flies in near-Earth orbit about 600 kilometers up.*

that in the ultraviolet keeps the burning—and lethal—solar ultraviolet at bay (although not entirely, as you can find out at the beach). X-ray radiation cannot penetrate, either, and the electrified layers of the Earth's upper atmosphere block several radio bands (standard "short-wave" communication in fact relies on bouncing signals off these layers from below). For all these reasons, beneficial as they may be in other respects, we are unable to study a huge variety of high- and low-energy processes from the ground. Even the optical region of the spectrum is compromised by some absorption and by the ever-present twinkling.

If we want further advances in our study of the Universe, we have no choice but to go above the atmosphere into space, where we can get an unobstructed view. Many rocket flights carrying diverse instruments have been made since the 1950s. A rocket payload in a ballistic trajectory affords only a few minutes' look, however. High-flying balloons have been used as observing platforms, too, but obviously these must swim in some residual atmosphere. The only serious way of observing from space is by satellite. The first truly astronomical satellite (as opposed to those that studied geophysical processes) was launched in the 1960s. Many more followed, and the instruments aboard them have opened up most of the hidden electromagnetic spectrum.

A few examples show the variety. The remarkable, unsung mechanized hero of the 1970s and 1980s was a 45-cm telescope called the *International Ultraviolet Explorer (IUE)*. It was launched in 1978 with a three-year design lifetime to observe all manner of celestial bodies in the ultraviolet below the atmospheric cutoff, in the region between 1000 and 3000 Å. After 19 productive years, the *IUE* finally was shut down. The public knew little about it because it was designed only to observe ultraviolet spectra and could not produce spectacular images. Without any question, however, it generated a revolution in astronomy, allowing us to witness physical processes at work that simply cannot be seen from the ground. Such ultraviolet imaging has been impressively done with *ASTRO,* an observatory that flew for a week aboard the space shuttle *Columbia* in May 1991.

X-rays reveal the locations of very high temperature gases and the sites of other extremely energetic processes. The first effective X-ray telescope, *Uhuru* (Swahili for "freedom"), flew in 1970 and immediately began finding sources of this energetic radiation. It was surpassed by *HEAO2 (High Energy Astronomical Observatory 2,* dubbed *Einstein*), which flew in 1978 and could image the sky. That satellite has now been superseded by *ROSAT (Röntgen-Satellite),* which is operated by the European Space Agency and NASA, and the *Compton Gamma Ray Observatory* is taking us to even higher

ASTRO, *the ultraviolet and X-ray observatory, glides above the Earth as astronomers observe from the shuttle bay.*

energies. In the other wavelength direction, the *Infrared Astronomical Satellite (IRAS)* flew in 1983. It imaged the entire sky three times in four wavelength bands, compiled infrared spectra, and has provided invaluable information on cool stars and the chilling depths of space. Its work was taken up in 1997 by the *Infrared Space Observatory*.

We are now immersed in discoveries by the *Hubble Space Telescope,* the *HST*. Begun in 1979 and finally launched in April 1990, it is returning to Earth stunning views of the heavens. Although far from the largest telescope in the world, it is surely the most sophisticated, and without any question it is in the best observing site. The *HST* (designed in the Ritchey–Chrétien configuration) carries a variety of imaging, photometric, and spectroscopic instruments. Its wide-field planetary camera has a theoretical resolution of 0.1 seconds of arc, several times better than can ordinarily be achieved from Earth, and the faint-object camera a phenomenal 0.043 seconds. The much-publicized problem with its 2.4-meter mirror (mistakenly ground to the wrong figure) initially compromised the instrument's resolution and efficiency, but the repair mission that took place in 1994 and the subsequent upgrade that added new instruments made the *Hubble* a spectacularly successful telescope, as shown in the pages to follow.

MEANWHILE, BACK ON THE GROUND . . .

Space astronomy does not supplant ground-based research but supplements it; the two work in superb cooperation. The developments taking place on Earth are as exciting as those hundreds or thousands of kilometers above us. For example, look at the Keck 10-meters or the battery 8-meters, that will have vastly greater light-gathering power than does the *HST*. More exciting perhaps is our growing ability to overcome the deleterious effects of our shimmering atmosphere to produce remarkably high resolution. Before long, telescopes on the ground will be able to produce images the like of—or even better than—*HST,* although they obviously will never rival its ultraviolet capability (unless we destroy *all* the ozone, a horrifying thought).

There are several ways to achieve that great goal of astronomy—high resolution—from the Earth. The history of interferometry goes back to 1913. In its simplest application, two slits are used to observe the diffraction pattern of a star. Since a star is not really a point, we get in effect a smeared pattern of fringes that allow the determination of the angular

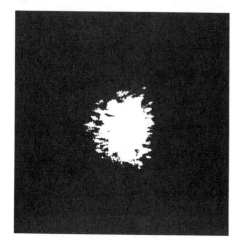

A speckle image of the star Capella. The star is angularly very small, its image heavily smeared by the turbulent air, but the exposure time is so short that we can actually see the myriad individual shifting images.

diameter. In the first such experiment, A. A. Michelson placed a 10-foot steel beam with mirrors at each end across the top of the 100-inch telescope on Mount Wilson. Two more mirrors reflected the light onto the objective, where the rays were mixed and the fringe pattern observed by eye. Michelson and others reached resolving powers of a few hundredths of a second of arc, allowing them to determine the angular diameters of some stars. Modern systems, which use photometry and computer processing, do even better and permit the observation of much fainter sources. Another method, called speckle interferometry, uses ultrashort exposures of stars and can freeze atmospheric seeing patterns, enabling us to record hundreds of tiny stellar images within the big seeing disk. Computer processing of several such images also allows the measurement of angular diameters. The technique is especially useful for the separation of stars that are so close together that they cannot be resolved by ordinary observations. Even more successful are optical versions of radio interferometers run by different groups that can achieve resolutions in the thousandths of seconds of arc.

The most exciting method uses a technique called adaptive optics. We equip the telescope mirror with small motors that can deform its surface. The twinkling of the stars is monitored through the telescope and the shape of the mirror surface quickly changed to compensate and to detwinkle the incoming light waves. In principle, we can then achieve the theoretical resolving power of the telescope. Resolutions of a few tenths of a second of arc—not much greater than that produced by the *HST*—have been achieved, and that is only the start. The goal of these techniques is to image the stars themselves, to see what is on them, even to look for their planets. Success should not be too long in coming.

JOIN IN

Astronomy is a remarkably accessible science—all you need do is go out and look. Furthermore, the chief astronomical instrument, the telescope, is an extremely simple affair that anyone can own and work with (not many amateur physicists have particle accelerators). There are many stars and few professionals, and it takes only some practice and a little dedication for the true amateur to make a real contribution to the celestial science.

The amateur can work at many different levels of interest and sophistication. There are groups of devoted observers who use only their eyes to track meteor showers across the sky. (Meteors are generally caused by fluffy

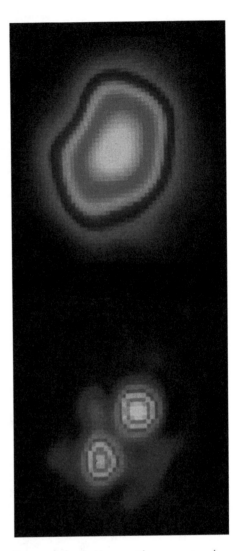

Two views of a double star with components only 0.38 seconds of arc apart; the one on top was taken under very good seeing conditions (0.8 seconds of arc) with the 3.6-meter New Technology Telescope at the European Southern Observatory. The lower image shows the salutary effect of switching on the adaptive optics, which produce a seeing disk only 0.22 seconds of arc wide.

dust thrown from comets that enters and burns up in the atmosphere. If the Earth passes close to a comet's orbit we see a brief shower as thousands of the little bits of matter plunge into the upper air.)

The first level of optical aid is a pair of binoculars. It is remarkable what can be seen with them, especially if you use a tripod to hold them steady. Binoculars are described by a pair of numbers, for example, 7 × 35. The first gives the power, the second the aperture in millimeters. For nighttime use, 7 × 50 is excellent, and bigger ones like 11 × 80 will provide a stunning wide-angle view of the heavens. Jupiter's little moons, the larger craters on the Moon, and dozens of star clusters will jump out at you.

To see the sights at their best, though, you must get a telescope. The problem is the huge choice. Although modern refractors deliver superb images, for the price you are always better off with a reflector. The minimum usable mirror diameter is about 75 mm. If constructed in the classic Newtonian design, the costs are modest and the views nearly inexhaustible. Obviously, the bigger the objective the better. Simple and reasonably inexpensive instruments are available up into the 60-cm range. At a more sophisticated, and expensive, level, you might opt for your own Schmidt (or "catadioptric") system. These use a Cassegrain design, in which a long effective focal length, affording high power, is folded into a short tube. They are therefore easy to transport into a dark site, a must for deep-sky viewing. All amateur instruments come with a variety of eyepieces, those of longer focal length for a wide field of view, those of shorter to see fine detail. Contrary to expectation, the better views are usually obtained with lower powers. The highest you can ordinarily use is about 50 power per inch of aperture, and even that is good only on exquisite nights.

Getting started requires more than just purchasing a telescope. Astronomical observing, at any level, is like any other endeavor: It is a learned skill that requires practice. Use your new telescope first in the daytime to look at distant trees and to learn how to use (and to align) the low-power finder. You will then become aware of one of the astronomical telescope's most confounding properties: The view is inverted. You move the instrument to the right, and everything appears to move to the right instead of the left. Just get used to it. Before long the Moon will look abnormal if presented to you right side up.

When you take your telescope out at night you will discover another problem. It magnifies not only the angular size of an object but also the apparent speed of the Earth's (or the sky's) rotation. You have your first object centered, and before you know it, it is gone. To follow it you have to move the telescope constantly, but in the "wrong" direction! Patience, you *do* get used to it. If you have paid enough, you may have a sidereal clock drive that will follow a star for you. But only, of course, if you have remembered to align the telescope's polar axis *on the pole.* That means a latitude adjustment and another at night to point the axis north (or south if you live in the southern hemisphere). A great deal of help—and fun—can be had from your local amateur astronomy club.

If you want to drive from Pocatello, Idaho, to Mobile, Alabama, you do not just jump into the car—you get a map. Similarly, maps help you find your way around the sky. Like terrestrial charts, sky maps come in all sizes and varieties. At first, you want one that will show you the bright stars and constellations so that you can get a good overview of the heavens. Then, when you are familiar with the basic layout, move up to more detailed charts that show all the naked-eye stars along with a variety of interesting celestial objects, or even to ones that show stars much fainter than the eye can see. You can then star-hop from a bright, easily locatable star to fainter ones visible only in your telescope and on to, say, a faint galaxy.

Numerous publications are available that can help enormously in guiding you around the sky. Many carry ads that will aid in finding the instrument of your dreams and the charts to go with them, and some list astronomy clubs that will help you find fellow enthusiasts. You might also discover that you wish to enter the ranks of research. Hundreds of comet-hunters scan the skies every evening and morning for these elusive balls of fuzz. The large majority of comets are found by amateurs who wander the open sky, not by the professionals who sit in darkened domes with only video screens for company. Thousands more amateurs monitor the myriad stars that vary in their light, working under the auspices of national groups (such as the American Association of Variable Star Observers), and for generations have provided valuable data to the world's professionals.

Whatever the level chosen, now we are prepared to go outside and look, to roam through the Universe with our eyes and our telescopes, and to *learn.*

Whether admiring Orion, peering through our binoculars, guiding our Newtonians, or controlling our 4-meters, we are all astronomers.

The Andromeda Galaxy, 750,000 parsecs away, is a system like our own.

Chapter Three

THE DISCOVERY OF REALITY

WHERE—AND WHAT—ARE THE STARS?

You are deceived when you look at the nighttime sky. Some stars are bright, mostly because they are close; others that look dim are incredibly luminous, but happen to be very far away. How do we sort them out and learn their true natures? The story involves the measurement of distance, the detection of motion, and the analysis of stellar light. We begin with the distance to the Sun, then take a giant leap into interstellar space.

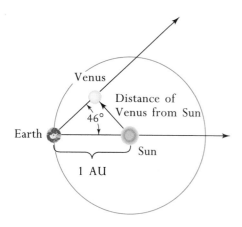

Venus' maximum angle from the Sun is 46°. From the right triangle, Copernicus found its distance from the Sun to be 0.7 times that of the Earth (the astronomical unit).

DISTANCES AND MOTIONS IN THE SOLAR SYSTEM

The problem of the solar distance, a fundamental quantity now called the astronomical unit, or AU, was attacked by the classical Greek astronomers. By Ptolemy's time there was at least an estimate, however erroneous, of 20 times the distance to the Moon and 600 times the diameter of the Earth. There the matter lay for 1700 years, until Copernicus (his proof supplied later by Galileo) established the concept that the Sun was indeed at the center of the solar system. With that knowledge, he could measure the relative sizes of the planets' orbits from their positions in the sky.

Copernicus' theory had a major flaw. Because he assumed circular orbits, as had the Greeks before him, the planets were never quite where predicted. Something had to be wrong. The revelation of the real structure of the solar system is the product of the awesome labor of two men, Tycho Brahe (1546–1601) and Johannes Kepler (1571–1630). Tycho had established an observatory on the Danish island of Hveen, and with meticulously graduated circles made careful measurements of the positions of stars and planets. Kepler worked with him in Prague in the last year of his life, and inherited the precious data. He adopted a revolutionary procedure. He made no presumption about orbital shapes, but used the data to find their natures. The result was a set of three laws that govern motions within the solar system.

In 1609, after eight years of crushing calculations in which he compared his orbital predictions with Tycho's positions of Mars, Kepler produced his first two laws. The first was astonishing: A planetary orbit is not a circle, but an *ellipse* with the Sun at one focus. The size of an ellipse is characterized by half the major axis, the semimajor axis, a. The astronomical unit is properly defined as the semimajor axis of the terrestrial orbit.

The second law governs the speed at which the planets move along their paths. Kepler found that the line connecting the planet to the Sun, called the radius vector, sweeps out equal areas in equal times, which means that a planet's orbital speed changes with its distance from the Sun.

The third law took another 10 years to uncover. This one, his "harmonic law," relates the orbits of the planets to the Earth and to one another: The squares of the planetary periods in years are equal to the cubes of their semimajor axes in AU, or $P^2 = a^3$. Jupiter, for example, takes 12 years to go around its orbital path. Twelve squared is 144, the cube root of which is 5.2 AU, the planet's distance from the Sun.

Kepler's three laws are empirical, derived from observation and calculation. Kepler had no idea why they worked. They were explained in the late 1600s by Isaac Newton, the extraordinary mathematician and thinker who established the now-familiar laws of motion:

1. A body in motion stays in motion unless acted upon by an outside force;
2. Force equals mass times acceleration ($F = MA$);
3. For every action there is an equal and opposite reaction.

The second law is the most important to us here. A force is anything that can cause an acceleration, or a change in velocity, where velocity is defined as a body's speed and direction. Mass is commonly thought of as the amount of matter within a body, but since $F = MA$, it can also be defined and measured as F/A, the degree of force needed to accomplish a particular acceleration (or the acceleration acquired by application of a specific force). It is measured in kilograms, kg.

Newton followed these with his famous law of gravity, which states that the same force that makes the apocryphal apple fall to the ground makes the Moon go around the Earth. The Moon is actually falling toward us but cannot reach us because it is also moving in a direction perpendicular to the line connecting it to the Earth. The acceleration of a falling body, g, is proportional to the mass of the Earth divided by the distance to the Earth's center squared (since the Earth attracts as if all the mass were concentrated at a central point), or

$$g = GM_{Earth}/R^2$$

where G, the gravitational constant, is determined in the laboratory. The acceleration can be generalized to any pair of attracting bodies, and since $F = MA$, the force between them, F, is written as

$$F = GM_1M_2/R^2$$

where M_1 and M_2 are the two masses and R is the separation of their centers.

These laws, together with his invention (simultaneously with Leibniz in Germany) of calculus, allowed Newton to derive Kepler's laws from theory in an elegant and generalized form of which Kepler's empirical versions were special cases. He found first that orbits involving two mutually gravitating bodies can be *any* of the conic sections, not only ellipses. An orbit may also be a parabola or a hyperbola, along which the orbiting body

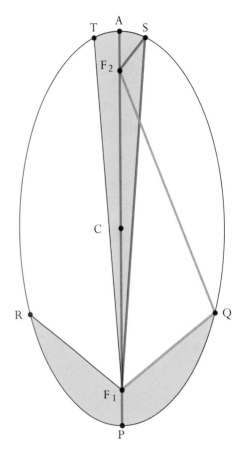

The sum of the distances of all points on an ellipse to each of the two foci, F_1 and F_2, is a constant: $F_1Q + F_2Q = F_1S + F_2S$, etc. The size of the ellipse is defined by half the major axis (the semimajor axis), CP, also called a, and the shape is given by its eccentricity, e, CF_1/a. In a planetary orbit, the Sun is at one of the foci. As the planet moves, the radius vector that connects the planet and Sun must sweep out equal areas in equal times, so the planet moves from Q to R in the same interval of time in which it moves from S to T. The planet moves fastest at perihelion, P, and slowest at aphelion, A. (Planetary orbits are actually much more circular than represented here.)

will escape. Kepler's second law is generalized by a concept called the conservation of angular momentum. Tie a rock to a string and whirl it about your head. The angular momentum is defined as the mass of the rock times its velocity times the length of the string. Once established, angular momentum will not change unless an outside force is applied: If you shorten the string, the rock speeds up. A skater bringing in his arms almost magically whirls faster. The conservation of angular momentum makes planets speed up as they approach the Sun on their elliptical orbits, and slow down as they recede.

Since orbits are determined by gravity, and gravity involves mass, Kepler's third law must somehow include mass as well as period and distance. In his generalization, Newton derived one of the most important relations in all of astronomy:

$$P^2 = \frac{4\pi^2}{G} \frac{a^3}{(M_{Sun} + M_{planet})}$$

The orbital parameters are now divorced from those of the Earth: The period *(P)* is measured in seconds, the semimajor axis *(a)* in meters. Kepler's original law lies buried within. Because the masses of the planets are so much less than the mass of the Sun, the sum $(M_{Sun} + M_{planet})$ is effectively a constant (at least to the accuracy of Tycho's naked-eye observations). If the equation as applied to any planet is divided by that for the Earth, the constants disappear, and $P^2 = a^3$, where *P* is again in years and *a* is in AU—just what was found by Kepler.

Newton's and Kepler's laws have far-reaching ramifications. Immediately, they allowed astronomers to determine accurately the relative separations between the various bodies of the solar system at any time in terms of the astronomical unit. A knowledge of the distance in kilometers to any of the planets serves to calibrate the AU and yields the distance to the Sun. The distances in kilometers to all the planets are then automatically known as well. The first such accurate measure was made more than 300 years ago by Giovanni Cassini, using his own observations and those made by the French astronomer Jean Richer. The two men took measurements from Paris and French Guiana respectively and determined the distance of the planet Mars from its terrestrial parallax (in this context, the shift in position as viewed from opposite sides of the Earth). During the nineteenth and twentieth centuries, the measure of solar distance was improved by observing the parallaxes and distances of close-passing asteroids, small bodies that ordinarily lie between the orbits of Mars and Jupiter. The modern value of the AU (1.459×10^8 km) is found by observing Venus with radar. We simply send a radio signal to the planet, find the interval of time between transmission and reception of the returned signal (allowing for slowing by refraction in the terrestrial and Venusian atmospheres), multiply by the speed of light, and divide by two. The result of all this effort, begun so long ago, is the distance to one very important star.

THE DISTANCES AND MOTIONS OF THE STARS

The Galaxy is a churning, rotating mass. All the stars are in orbit about the galactic center, and all the orbits are at least slightly different from one another, some wildly so. From our home, the Sun, we would expect to see the stars constantly moving past one another. Think of a body, say an automobile, passing by you as you stand still. The vehicle has a speed (read on the speedometer) and a direction of motion that combine into its space velocity, v_s. The car is most likely moving at an angle relative to your line of sight. Its speed can then be resolved into two components, the radial velocity, v_r, along the line of sight and the tangential velocity, v_t, perpendicular to it. The tangential movement causes the car to appear to move across your line of sight at a certain angular rate of so many degrees per second, the "proper motion." The proper motion depends on both the tangential velocity and the distance: A car moving along the street where you stand whizzes by, whereas one seen on a distant highway seems to crawl along the horizon.

Instead of cars, think of stars: The principles are exactly the same. Measurement of proper motion, μ, and distance, d, allows the calculation of v_t from simple trigonometry. The radial velocity, v_r, can be found by the Doppler formula. With these quantities known, we can then determine the actual direction and speed of the star, the space velocity, where $v_s{}^2 = v_r{}^2 + v_t{}^2$ from the Pythagorean theorem.

The first of these quantities to be observed were the proper motions. They cause changes in right ascension and declination (apart from those produced by precession), a discovery made in 1718 by Edmund Halley, who did far more than invent a constellation and get his name attached to a comet. They are now known for tens of thousands of stars. Because of the great distances involved, most of the movements are small, only a few tenths, hundredths, or even thousandths of a second of arc per year, and are sensible only with careful telescopic measurements. Nevertheless, given enough time they will cause great changes in constellation patterns.

Aristotle, Copernicus, and others all knew that if the Earth indeed orbits the Sun, the stars must exhibit annual shifts or parallaxes. Moreover,

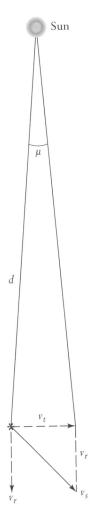

A star moves relative to the Sun at a space velocity v_s. The motion is resolved into two mutually perpendicular components, the radial velocity, v_r, and the tangential velocity, v_t. Over a given interval of time, the star will move through a proper motion, μ, which depends on both v_t and distance, d.

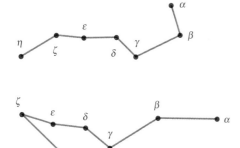

A quarter-million years from now, the Big Dipper (top) will be seen as the figure on the bottom.

As recorded in photographs made 32 years apart, the double star 61 Cygni sails through space. Note its altered position with respect to the stars at upper right (top).

since the sizes of the shifts must be inversely proportional to the distances, the distances can be measured. The stars are so far away, however, and the parallaxes so tiny, that the quest for stellar distance was not successful until 1838, when Friedrich Wilhelm Bessel turned his telescope and his trained eye to the star 61 Cygni. In his time this star, actually a double, had the highest known proper motion, a whopping 5 seconds of arc per year (it is now number two to 10th-magnitude Barnard's Star in Ophiuchus, which is clipping along at double that rate). It therefore had a good chance of being one of the closest stars, one with a relatively high parallax, and worthy of careful examination.

After allowing for the effects of proper motion, Bessel found that over the course of the year the components of 61 Cyg shifted back and forth through an angle of ⅔ of a second of arc. This small angle, $\frac{1}{5400}$ of a degree, is the angular size of the width of your thumb at a distance of 6 km. The astronomical parallax, p, is defined as ½ the shift angle, and so for 61 Cyg it is ⅓″. The distance unit used in professional astronomy is the parsec, pc, defined simply as $1/p$, which renders the star 3 pc away (the improved modern value is 3.4 pc). A parsec is the distance at which the Earth's orbital radius would subtend an angle of 1 second of arc, which is 206,265 AU (the number of seconds of arc per radian, an almost magic number in astronomy). The star nearest to the Earth (other than the Sun), however, is not 61 Cyg, but a faint, reddish 11th-magnitude companion to α Centauri known as Proxima Cen, which has a huge parallax of 0.76 seconds of arc and a distance of 1.3 pc, or 270,000 AU. If the Earth's orbit is the size of a golf ball, α Cen is some 5 km away, and the Sun a mere 0.1 mm across—no wonder that stars do not collide! Once we have the distances, transverse velocities can easily be calculated from proper motions. Now we can know something of the relative speeds of the stars, and we find them to be typically in the neighborhood of a few tens of kilometers per second.

The more famous unit, the light-year, is quite useful for the comprehension of stellar distances. The light year, l-y, the distance that a ray of light travels in one year at a speed of 300,000 km/sec, is equal to 10^{13} km or 3.26 pc. The light you see from α Cen left the star *four years ago*. Of course, this star just begins our journey into space. Most of those you see in the nighttime sky are dozens, hundreds, even thousands of parsecs and light-years away. Their assembly, the Galaxy, is well over 30,000 parsecs—100,000 light-years—across. And we can take images of individual stars in other galaxies that lie *millions* of parsecs away.

Parallaxes are very difficult to measure, and their determinations require years of observation. The minimum detectable shift has long been around 0.01 seconds of arc, limiting the accurate application of the tech-

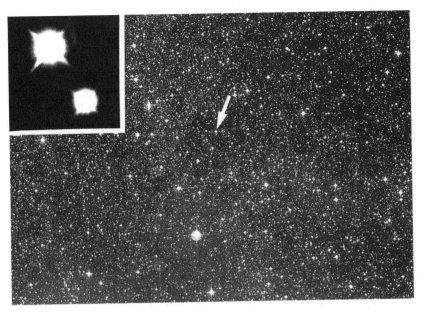

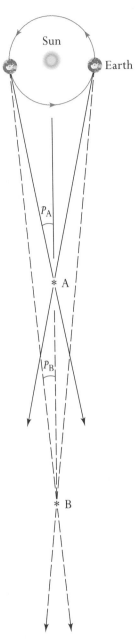

Proxima Centauri (arrow) is the closest star to Earth. It is gravitationally bound to α Centauri (inset), a double star too far away from Proxima to be seen in the main picture.

nique to distances closer than about 50 to 75 pc. However, the frontier is being pushed back both from the ground and from space. The *Hipparcos* satellite has measured parallaxes and proper motions for hundreds of thousands of stars with errors of 0.001 seconds or better, and good distances up to 200 pc or so. Even better measures will surely follow.

TRUE BRIGHTNESS

We have broken through the celestial sphere. Now that we know that stars reside in the three-dimensional domain and we can plot their locations, we must expand the definition of magnitude. That measured from the Earth on the basis of direct observation by eye or telescope is the apparent magnitude, *m*. We now need a magnitude that removes the effect of distance and is a measure of true luminosity or energy output. This quantity is the absolute magnitude, *M*, defined as the apparent magnitude the star would have were it at an arbitrary standard distance of 10 pc, 32.6 l-y. Absolute magnitudes are easy to calculate once the difficult matter of distance is

A pair of stars, B twice as distant as A, are observed in the ecliptic plane. Each will seem to swing back and forth over a period of a year. The parallax, half the total shift, is the angular size of the Earth's semimajor axis as viewed from the star. The parallax (P) of B is half that of A.

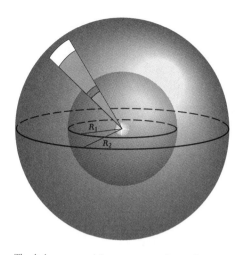

The light generated by a star spreads itself out over a sphere of radius R_1. As the size of the sphere increases, the amount of radiation per unit area becomes diluted by $1/R^2$. Double the distance to R_2, and the source looks a quarter as bright.

solved. The brightness of a light source depends on the inverse square of its distance; halve the distance and the source is four times as bright. Select a sixth-magnitude star 100 pc away. If we could move the star in to 10 pc, it would become 100 times, or five magnitudes, brighter, so that M would be $+1$. Any apparent magnitude can be converted to absolute through a simple formula, the famed magnitude equation,

$$M = m + 5 - 5 \log d$$

where d is in parsecs. Mark this little equation well; it will play a powerful role as we proceed.

If we could line up all the stars at 10 pc the sky would look very different. Start with the Sun, whose distance from us is 1 AU or $1/206{,}265$ pc. The Sun's apparent magnitude is -26.74, and when the numbers are entered into the equation, the absolute magnitude is $+4.83$, so faint that it could not be seen under full moonlight. Then look at Vega. At a distance of 7.5 pc, it still has m equal to 0 and is nearly 50 times more luminous than the Sun. In stellar terms, however, Vega is not extraordinary. Just east of it, in Cygnus, lies the great star Deneb, $m = 1.25$, distance 500 pc. Bring it in to 10 pc and M jumps to -7.2, rendering it nearly 1400 times brighter than Vega and *65,000 times the visual solar luminosity!* If the Earth were in orbit about such a beast, we would have to be 250 AU away, six times the distance of Pluto from the Sun, for humanity to survive. And even that is not the record. A handful of stars can be found that are as bright as $M = -10$, another 13 times brighter than Deneb.

From these heights, plunge to the depths. Recall 11th-magnitude Proxima Centauri, the closest star, only 1.3 pc away, yet visible in a modest telescope. Take it out to 10 pc and the brightness drops nearly 60 times to magnitude 15.5, that is, 10.6 magnitudes or 18,000 times *fainter* than the Sun. If we wished our day to be comparably bright under Proxima's feeble glow, we would have to cuddle in a distance of only 0.008 AU, *1.5 times the radius of the Sun.*

Amazingly, Proxima is downright bright compared with a faint star in Cepheus called LHS 2924, absolute magnitude 20, which is visually another 70 times fainter. In order even to see this meager star with the naked eye, it would have to be a mere 0.02 pc, 4000 AU, away, about 100 times farther than Pluto. If it were at Pluto's distance, LHS 2924 would appear only as bright to us as Venus; compare it with Deneb, already first magnitude at a distance of 500 pc. Absolute stellar brightnesses range over 30 magnitudes, an astounding factor of a *trillion,* from roughly a million times brighter than our Sun to a million times fainter.

COLOR AND TEMPERATURE

People are often astonished to find that stars show colors, some quite vivid, evident in photographs and indeed simply by eye on a clear night: Betelgeuse and Antares are rather red, Arcturus orange, the Sun and Capella yellowish, Vega and Sirius quite white, and a few are even somewhat bluish. But color represents more than just a handsome sight: It is a response to stellar temperature.

To a reasonable approximation, a star behaves something like an ideal radiator called a blackbody, a surface that absorbs all the radiation that falls upon it. It reflects none, hence the term black. Absorption of radiant energy would by itself increase the temperature of the body. In order to maintain equilibrium, it must *emit* just as much energy as it receives, and consequently can be very bright to the eye. The spectrum emitted by a blackbody is a continuum in which the different colors and wavelengths run together

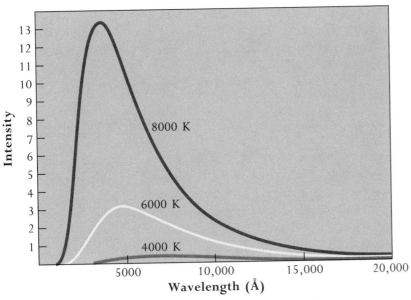

The spectra of blackbodies display a slow rise to a maximum as the wavelength decreases, followed by a rapid fall. Here are the spectral curves of three such bodies (of the same size), radiating at 4000 K, 6000 K, and 8000 K. The peak of the emission, λ_{max}, shifts to shorter wavelengths as temperature is increased, making the hotter body bluer. In addition, the amount of radiation at every wavelength is increased. The total luminosity, the integrated area under the curve, is proportional to the temperature raised to the fourth power.

with no breaks or gaps. The brightness of the spectrum varies with wavelength in a very characteristic way. As λ decreases from high values, the radiated energy steadily increases to a maximum, then drops suddenly to near zero. As the temperature, T, increases, the blackbody curve keeps the same relative shape but increases in height and expands toward lower wavelengths. At only a few degrees above absolute zero ($-273°C$), it radiates only in the radio spectrum. In the hundreds of degrees Kelvin (or Kelvins, K, centigrade degrees above absolute zero), it emits in the infrared as well, but is still invisible to the eye. But as T tops a thousand or so degrees K, it begins to glow in the optical (visible) spectrum, steadily increasing the amount of blue light relative to yellow and yellow light relative to red. The effect is easily seen in a toaster or oven element, which changes color from red to orange as it heats. Theory shows that the wavelength at which the emission is at a maximum is given by the Wien law:

$$\lambda_{max}(\text{Å}) = 2.9 \times 10^{7}/T$$

Measurement of the spectrum allows determination of the temperature.

The amount of radiation released by a blackbody per square meter of surface per second, a quantity called the total flux, F, is proportional to the fourth power of the temperature. That is, in units of watts per square meter per second ($\text{m}^{-2}\ \text{sec}^{-1}$),

$$F = \sigma T^{4}$$

(an equation called the Stefan–Boltzmann law), where σ is a constant of proportionality determinable in the laboratory. One body twice as hot as another will be 16 times as bright per unit area of surface. The gross energy output of the blackbody, called its luminosity, L, must then be equal to the body's surface area times the flux. If the blackbody is a sphere, whether a star or a bowling ball, the surface area is $4\pi R^{2}$, and $L = 4\pi R^{2}\sigma T^{4}$, certainly one of the most important relations in astrophysics. This enormously useful equation allows the derivation of L, R, or T if the other two are known.

Now the origin of the stellar colors is clear. They relate to different surface temperatures through the Wien blackbody law: Red stars are relatively cool, near 3000 K; yellow ones are closer to 6000 K; white are about 10,000 K; and the blue stars are hotter yet, perhaps 20,000 K or greater. Moreover, color is a quantifiable stellar property that allows the derivation, or at least the estimation, of temperature.

Earlier we expanded the definition of magnitude into space to allow for distance; now we expand the definition into different parts of the spectrum. The apparent magnitude of a star, m, as described in Chapter 1 (or as mimicked with photoelectric photometers or CCDs), is actually called the *visual* apparent magnitude, m_V, or just V, and it corresponds principally to the yellow part of the spectrum (technically at 5480 Å) where our eyesight is most sensitive. The magnitudes in Appendix 3 are all V. Astronomers also measure magnitudes in the blue part of the spectrum (where the first photographic plates were most sensitive, roughly corresponding to 4500 Å), called B.

The difference between the two magnitudes, $B - V$, is the color index. The color indices of white stars are near zero, those of blue stars are slightly negative (because the star is brighter at B and has a lower magnitude number), and those of red stars can be quite positive. At B, brilliant Betelgeuse is third magnitude! Magnitude systems can be elaborate; in addition to V and B, astronomers have added an ultraviolet magnitude U (3500 Å), which gives another color index, and several more that extend into the red and infrared.

Measurement of the color index gives the ratio of the luminosities at two wavelengths on the blackbody (or stellar-energy distribution) curve, which then tells the temperature. However, B and V by themselves provide limited information, because they are appropriate to specific wavelength bands. More physically meaningful is a quantity called the bolometric magnitude, m_{bol}, which corresponds to *all* the radiation of the star, that integrated over all wavelengths. It can be calculated if we know the stellar temperature, which gives us the form of the blackbody curve (or the true distribution of stellar energy), and the amount of radiation not included in V. For convenience, bolometric magnitudes are scaled to V for yellow-white stars. Consequently, for medium-temperature stars like the Sun, Vega, or even Arcturus, which radiate much of their energy in the visual spectrum, the bolometric magnitude does not differ much from V. For extremely hot or cool stars, which radiate profusely in the ultraviolet or infrared, the differences can be several magnitudes and are critically important for the determination of physical stellar properties.

Now combine the effects of distance and temperature. Absolute magnitudes can be expressed in any color. Apply the magnitude equation to V and you get absolute visual magnitude, M_V; apply it to B to find M_B, or to m_{bol} for M_{bol}. The color index remains the same whether using apparent or absolute magnitudes.

As useful as they are, however, absolute magnitudes tell us nothing about the energy output of a star in physical units. Return to the Sun. It is

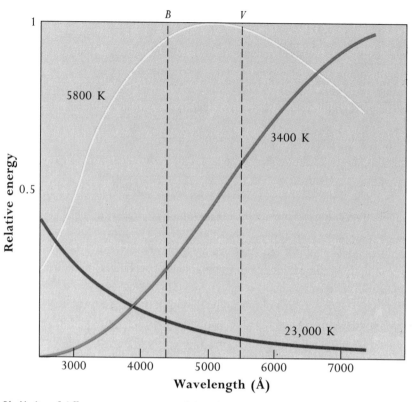

Blackbodies of different temperature are scaled to the same relative heights. The hottest (23,000 K) is brighter at B than at V, the coolest (3400 K) brighter at V. Since the magnitude goes down as the brightness goes up, the color index B − V must be negative for the hot blackbody and positive for the cool one. It is only slightly positive for the Sun at 5800 K.

a fairly simple matter to measure the solar constant, the rate at which the Sun heats the surface of the Earth, that is, the rate at which solar energy falls per square meter per second. Observation by Earth satellite, from which we can account for energy at all wavelengths, gives a rate of 1368 watts m^{-2}. Each square meter in a sphere with a radius R equal to 1 AU, or 1.5×10^{11} m, receives energy at this rate, so we multiply the solar constant by $4\pi R^2$ (the area of the sphere around the Sun with a radius of 1 AU) to find a total solar power of 3.8×10^{26} watts! Since a watt corresponds to a flow of 1 joule of energy per second, the Sun produces a phenomenal 4×10^{26} joules s^{-1}. And since a typical optical photon carries only 4×10^{-19} joules, something like 10^{45} photons emanate from the solar surface in that tiny fraction of time!

It is hard to have an appreciation for these numbers. Let your local power company run the Sun for a second and send you the bill. Typical energy rates are about $0.06 per kilowatt hour, which is 1.7×10^{-8} per watt second. In 1 second you will spend $4 \times 10^{26} \times 1.7 \times 10^{-8} = \$7 \times 10^{18}$. All you need do is write a check for the amount of the gross national product of the United States ($\$10^{12}$) for the next *7 million years!*

We are now in a position to calculate the solar temperature with precision. If $L = 4\pi R^2 \sigma T^4$, then T is the fourth root of $L/4\pi R^2 \sigma$. We already have L. From the Sun's distance and its measured angular diameter of 32 minutes of arc, its radius is easily found to be 7.0×10^8 m, or 108 times the radius of the Earth. The equation yields a temperature of 5780 K. Such a value is called an effective temperature, since it is the temperature that a blackbody of solar size would need to produce the observed solar luminosity. The surfaces of stars are not solid but semitransparent and gaseous, and as we will vividly see in Chapter 4, the temperature increases with depth. The result is that stellar spectra generally deviate (some considerably) from ideal blackbodies. Effective temperatures consequently provide an excellent means of intercomparison.

STELLAR SIGNATURES

Magnitudes and colors are important, but without spectra you cannot begin to understand the natures of the stars. Isaac Newton discovered that sunlight could be broken into its spectral colors. Then, in 1802, the English scientist William Wollaston found that the solar spectrum was cut by several dark gaps now called absorption lines. By 1815 Joseph von Fraunhofer had catalogued the wavelengths of more than 300 of them, assigning roman letters to the most prominent. Within a few decades laboratory investigations showed that individual lines were caused by specific kinds of atoms:

Fraunhofer's solar spectrum is crossed by numerous dark lines, each of which can be related to a particular kind of atom.

For example, those with wavelengths of 6563 Å and 4861 Å were associated with hydrogen, and a close pair in the yellow at 5801 Å with sodium.

When the nineteenth-century astronomers first turned their primitive visual spectroscopes on the stars, they were mightily puzzled by the spectral variety they saw. Only a few had absorption spectra like the Sun's. Some spectra were much simpler, but others appeared quite complicated, and the complexity correlated remarkably with color. By the late nineteenth century, hot white stars like Vega and Sirius were known to be dominated by a few powerful lines of hydrogen. Some blue ones had hydrogen and what was later determined to be helium. The orange-red stars showed little or no hydrogen but had strong bands now known to be

SOME PROMINENT SPECTRAL LINES

WAVELENGTH Å	FRAUNHOFER NAME[1]	ORIGIN[2]
6563	C	hydrogen (H)
5893	D	neutral sodium (Na I)
5876	. . .	neutral helium (He I); hot stars
5270	E	neutral iron (Fe I)
5167, 5173, 5184	b	neutral magnesium (Mg I)
4955	. . .	titanium oxide (TiO); cool stars
4861	F	hydrogen (H)
4686	. . .	ionized helium (He II); very hot stars
4384	d	Fe I
4300	G	CH molecule
4340	. . .	hydrogen (H)
4227	g	neutral calcium (Ca I)
4101	h	hydrogen (H)
3968	H	ionized calcium (Ca II)
3934	K	ionized calcium (Ca II)

[1] These letters were used by Joseph von Fraunhofer in the early 1800s to designate spectrum lines before they were chemically identified. The lines A and B, not listed, are produced by the Earth's atmosphere.

[2] "I" denotes the spectrum of a neutral atom, "II" that of the singly ionized form, and so on.

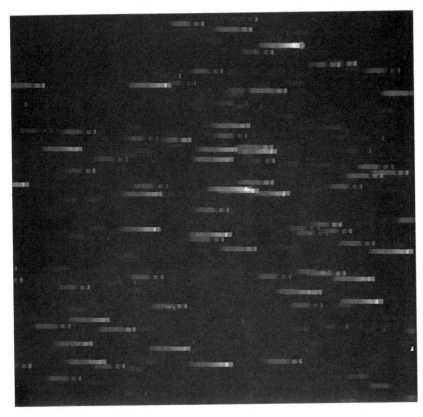

An objective-prism spectrogram showing the spectra of stars in the Hyades; note the variety. With practice it becomes easy to look at such a plate and classify dozens of stars quickly by eye.

caused by titanium oxide molecules, while the deep red ones displayed different kinds of bands produced by carbon compounds. Moreover, it looked very much as if stars of different temperatures had different chemical compositions.

The first step toward understanding in any new science is the classification of the data. The work was begun by several astronomers in the 1800s and culminated in a system developed around 1890 that is still in use today. E. C. Pickering, the director of the Harvard College Observatory, simply assigned letters to stars according to the strengths of the hydrogen lines, A for the strongest, B the next, down to O, the letters more or less correlating with color from white through yellow to red. The only glaring exception was class O, which followed red M and N even though its stars were blue-white. At about the same time, Pickering and his co-workers, notably Williamina P. Fleming, Antonia Maury, and Annie Jump Cannon, began the classification of stars in both celestial hemispheres with the aid of an

Annie Jump Cannon (1863–1941).

objective-prism spectrograph. This simple device uses a low-angle prism placed over the lens of a telescope, so that instead of recording the images of stars, we record those of the stars' spectra, allowing rapid classification of many at a time. It quickly became apparent that several of the original letters were either unneeded or were the result of flawed exposures, so C, D, E, and some other letters vanished. With the accumulation of a large number of spectra, Maury and Cannon realized that the gradations in the strengths of the spectrum lines were continuous only if class B were placed *before* A and O before B, resolving the color issue.

The final result is the spectral sequence of the seven basic stellar types, OBAFGKM. As the observations improved, Cannon found that the seven divisions were simply too coarse, so she decimalized them, making the A stars run from A0 at the hotter end to A9 at the cooler, followed by F0, and so on. In this system our Sun is classified as G2. The spectral classes of the 40 brightest stars are given in Appendix 3. By 1920, Cannon had almost singlehandedly classified 225,000 stars in the bible of early spectroscopy, the *Henry Draper Catalogue* (stars are commonly named by their HD numbers), and by 1940, 359,000 were available. It is not surprising that her name is one of the most revered in all astronomy.

THE SPECTRAL CLASSES

CLASS	CHARACTERISTIC SPECTRA	COLOR	COLOR INDEX	EFFECTIVE TEMPERATURE (K)	EXAMPLES
O	He II; He I	blue	−0.3	28,000–50,000	χ Per, ε Ori
B	He I; H	blue-white	−0.2	9900–28,000	Rigel, Spica
A	H	white	0.0	7400–9900	Vega, Sirius
F	metals; H	yellow-white	0.3	6000–7400	Procyon
G	Ca II; metals	yellow	0.7	4900–6000	Sun, α Cen A
K	Ca II; Ca I; other molecules	orange	1.2	3500–4900	Arcturus
M	TiO; other molecules; Ca I	orange-red	1.4	2000–3500	Betelgeuse
R[1]	CN; C_2	orange-red	1.7	3500–5400	. . .
S[2]	ZrO; other molecules	orange-red	1.7	2000–3500	R Cyg
N[1]	C_2	red	>2	1900–3500	R Lep

[1] Carbon stars

[2] Mild carbon stars

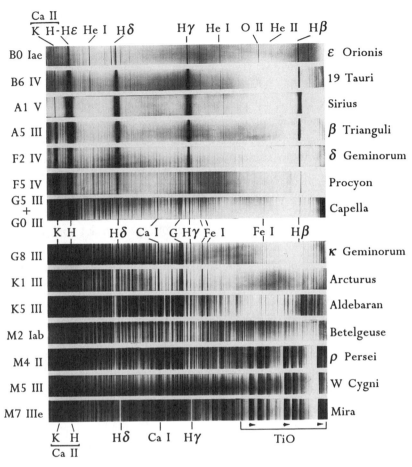

The classic spectral sequence observed. The strengths of the hydrogen lines rapidly fade upward and downward from type A. Several other lines are identified. Modern classifications of these stars are given on the left; roman numerals give luminosity classes.

DECODING THE MESSAGE

The different stellar spectra are caused by atoms in differing states of ionization and electronic excitation. The full explanation had to wait until the revelations of the 1920s showed us how the atom works. The basic model is a cloud of one or more negatively charged electrons surrounding a nucleus that consists of some combination of positively charged protons and neutral

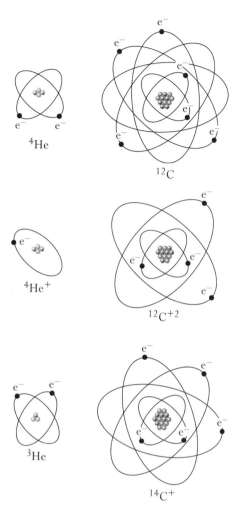

A helium atom (top left) *has two electrons*
orbiting a nucleus that consists of two protons and
two neutrons, and is therefore called ^{4}*He (the*
superscript—the atomic mass—counts the total
number of nuclear particles). Middle left: One
electron has been stripped from the atom, leaving
a positively charged ion He^+—*or more properly*
^{4}He^+. Bottom left *is the only other possible*
helium isotope, a rare form with only one neutron,
^{3}*He. At right are three carbon atoms, from top,*
^{12}*C,* ^{12}C^{+2} *(doubly ionized carbon, which has*
two electrons removed), and rare singly ionized
^{14}C^+. ^{13}*C is relatively common. Sizes of nuclei*
(in reality they are about 10^{-5} *the diameter*
of the electron orbits) are enormously exaggerated.

neutrons. The charges on the proton and electron are equal in amount, the mass of the proton is about equal to the mass of the neutron, and that of the electron is some 1800 times smaller.

Opposite charges attract, allowing the electrons to be tied to the nucleus. However, like charges repel. How then can the protons stick together? There are four fundamental forces of nature, invisible fields that act over a distance and control the Universe. By far the weakest, surprisingly enough, is gravity. It feels powerful only because there are so many atoms in the Earth to attract us. Stronger is the electromagnetic force, which controls electricity, magnetism, the attraction and repulsion of the different charges, and the creation of photons. Both these forces behave according to inverse square laws and operate over all space, the other two only over the short range of the nucleus itself. The "weak force" involves nuclear reactions; by far the strongest of the four is the attractive "strong" or nuclear force. If protons are separated, they will indeed fly apart, but if they are close enough, the strong force overwhelms the electromagnetic repulsion, and they stay locked together.

A specific chemical element is determined solely by its number of nuclear protons, a quantity called the atomic number: for example, hydrogen (H) has 1, helium (He) 2, carbon (C) 6, gold (Au) 79. A neutral atom has as many protons as electrons, but the electrons are tied fairly loosely, and if enough energy is supplied to the atom, one or more can be lost to produce positively charged ions. Carbon with one electron gone is C^+ and oxygen with three missing is O^{+3}. The sum of the number of neutrons and protons in the nucleus is the atomic mass, written as a superscript to the chemical symbol. Common helium, ^{4}He, has two of each, and carbon, ^{12}C, has six of each. As atomic number goes up, the neutron count tends to climb faster than the proton number, so that uranium, ^{238}U, has 92 protons but 146 neutrons. For any element the number of neutrons varies, producing one or more isotopes. Helium also comes with one neutron, called ^{3}He, and carbon comes in ^{12}C, ^{13}C, and ^{14}C varieties. Usually one isotope dominates over the others (there is only 1 atom of ^{3}He for every 100,000 of ^{4}He). The most common form is given in the periodic table of the elements.

Absorption lines are produced when an atom's (or ion's) electrons absorb energy from the continuum. Electrons bound to an atom are constrained to move in "orbits" of specific radii. (The analogy to planetary orbits is quite loose; in reality, the electrons behave as clouds of charge surrounding the nucleus.) The larger the radius, the higher the electron's energy. An electron can jump from a lower to a higher orbit if it absorbs a photon whose energy is exactly equal to the difference in energy between the two. The energy of a photon is equal to hν (Planck's constant times

The periodic table:

Key (legend):
- 14 — Atomic number
- Si — Symbol
- Silicon — Name
- 28 — Atomic weight (most common isotope)

1																	2
H Hydrogen 1																	**He** Helium 4
Li Lithium 7 (3)	**Be** Beryllium 9 (4)											**B** Boron 11 (5)	**C** Carbon 12 (6)	**N** Nitrogen 14 (7)	**O** Oxygen 16 (8)	**F** Fluorine 19 (9)	**Ne** Neon 20 (10)
Na Sodium 23 (11)	**Mg** Magnesium 24 (12)											**Al** Aluminum 27 (13)	**Si** Silicon 28 (14)	**P** Phosphorus 31 (15)	**S** Sulfur 32 (16)	**Cl** Chlorine 35 (17)	**Ar** Argon 40 (18)
K Potassium 39 (19)	**Ca** Calcium 40 (20)	**Sc** Scandium 45 (21)	**Ti** Titanium 48 (22)	**V** Vanadium 51 (23)	**Cr** Chromium 52 (24)	**Mn** Manganese 55 (25)	**Fe** Iron 56 (26)	**Co** Cobalt 59 (27)	**Ni** Nickel 58 (28)	**Cu** Copper 63 (29)	**Zn** Zinc 64 (30)	**Ga** Gallium 69 (31)	**Ge** Germanium 74 (32)	**As** Arsenic 75 (33)	**Se** Selenium 80 (34)	**Br** Bromine 79 (35)	**Kr** Krypton 84 (36)
Rb Rubidium 85 (37)	**Sr** Strontium 88 (38)	**Y** Yttrium 89 (39)	**Zr** Zirconium 90 (40)	**Nb** Niobium 93 (41)	**Mo** Molybdenum 98 (42)	**Tc** Technetium 99 (43)	**Ru** Ruthenium 102 (44)	**Rh** Rhodium 103 (45)	**Pd** Palladium 106 (46)	**Ag** Silver 107 (47)	**Cd** Cadmium 114 (48)	**In** Indium 115 (49)	**Sn** Tin 120 (50)	**Sb** Antimony 121 (51)	**Te** Tellurium 130 (52)	**I** Iodine 127 (53)	**Xe** Xenon 132 (54)
Cs Cesium 133 (55)	**Ba** Barium 138 (56)	**La** Lanthanum 139 (57)	**Hf** Hafnium 180 (72)	**Ta** Tantalum 181 (73)	**W** Tungsten 184 (74)	**Re** Rhenium 187 (75)	**Os** Osmium 192 (76)	**Ir** Iridium 193 (77)	**Pt** Platinum 195 (78)	**Au** Gold 197 (79)	**Hg** Mercury 202 (80)	**Tl** Thallium 205 (81)	**Pb** Lead 208 (82)	**Bi** Bismuth 209 (83)	**Po** Polonium 210 (84)	**At** Astatine 210 (85)	**Rn** Radon 222 (86)
Fr Francium 223 (87)	**Ra** Radium 226 (88)	**Ac** Actinium 227 (89)	104 261	105 262	106 263	107 262	108	109									

Metals | Nonmetals

Lanthanide Series

58	59	60	61	62	63	64	65	66	67	68	69	70	71
Ce Cerium 140	**Pr** Praseo-dymium 141	**Nd** Neodymium 142	**Pm** Promethium 145	**Sm** Samarium 152	**Eu** Europium 153	**Gd** Gadolinium 158	**Tb** Terbium 159	**Dy** Dysprosium 164	**Ho** Holmium 165	**Er** Erbium 166	**Tm** Thulium 169	**Yb** Ytterbium 174	**Lu** Lutetium 175
90	91	92	93	94	95	96	97	98	99	100	101	102	103
Th Thorium 232	**Pa** Protactinium 231	**U** Uranium 238	**Np** Neptunium 237	**Pu** Plutonium 242	**Am** Americium 243	**Cm** Curium 247	**Bk** Berkelium 249	**Cf** Californium 251	**Es** Einsteinium 254	**Fm** Fermium 253	**Md** Mendelevium 256	**No** Nobelium 254	**Lr** Lawrencium 257

Actinide Series

The periodic table lists all the elements in order of increasing atomic number. Vertical columns have similar chemical properties. The elements in the yellow boxes have been observed in stars.

frequency), which is also equal to hc/λ. Consequently, the energy difference between orbits defines the wavelength of the absorbed photon. Conversely, an electron will jump *down* between these two states by *emitting* a photon of the same wavelength, which produces a bright-line or emission spectrum.

You will see an absorption spectrum when a cooler, lower-density gas is placed in front of a hotter blackbody (or some similar source of continuous radiation). The emission spectrum is seen when a low-density gas is viewed by itself. Stellar absorption lines are created in a cooler, lower-density atmosphere at the star's surface that partially blocks the continuum emanating from the hotter, denser gases below.

To understand how the spectral sequence is created, look at simple hydrogen. The optical absorption (Balmer) lines arise from the second orbit: The orbit 2-to-3 transition is called Hα (either in absorption or emission), 2-to-4 is Hβ, and so on. The presence of the Balmer lines requires a sufficient number of electrons in the second orbit. But nature always seeks the lowest energy state, so the vast majority of atoms have their electrons in the lowest orbit. They are raised into orbit number two largely by collisions among atomic neighbors, but that requires the gas temperature and the particle speeds to be sufficiently high (even then, only a tiny number make it). In the cool atmosphere of an M star, the collisions are ineffectual, no electrons are preraised, and there are no hydrogen lines, even though the gas is 90 percent H. We first see the hydrogen lines in class K at about 4000 K, and they rapidly become stronger up about to type A at 10,000 K as more and more electrons are excited.

Then comes a surprise. Any hotter, and the collisions become so vigorous that some of the hydrogen is ionized. The result is that there are fewer neutral atoms and the hydrogen lines weaken, the effect exaggerated by the increasing opacity of the stellar gases, which makes the absorbing layers thinner. The energy levels that produce helium are much harder to excite than are those of hydrogen. Consequently, in order for them to be seen the temperature must be much higher. They are not found until class B, and ionized helium does not appear until nearly class O.

More complex atoms, the "metals"—the term is used here in its astronomical sense to mean any element heavier than helium—have more electrons and much more complex spectra. Nevertheless, they behave in similar ways. Neutral calcium is powerful at the low temperatures of the M stars. But calcium is much easier to ionize than hydrogen, so that as the temperature increases to only 5000 K or so, the neutral state gives way to the singly ionized form (with one electron stripped away). Furthermore, the ionized lines in the optical, Fraunhofer H and K, arise from the lowest orbit, so that no pre-excitation is needed. Therefore the solar Ca II lines (II denotes the spectrum of singly ionized calcium, Ca^+, III that of Ca^{+2}, and so on) dominate hydrogen, even though there is only one calcium atom to every 10^5 hydrogen atoms. Eventually Ca II weakens, as the temperature becomes great enough to knock yet another electron off to produce the doubly ionized form, which is followed by triple ionization. The lines of the highly ionized heavier elements tend to be in the ultraviolet, and so the optical spectra of the hotter stars become simpler, dominated by H and He. By contrast, at the lowest temperatures atoms can actually combine to form molecules, which produce extremely complex banded spectra. There is a

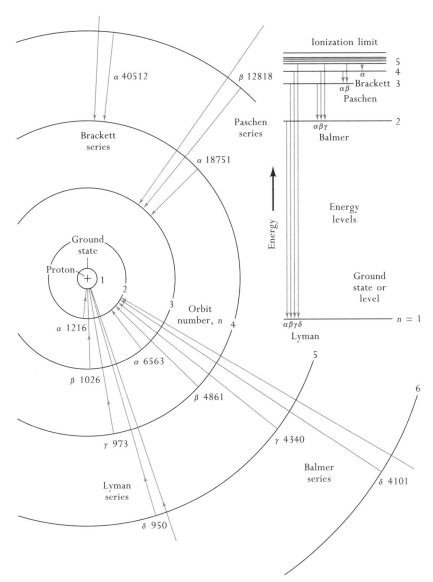

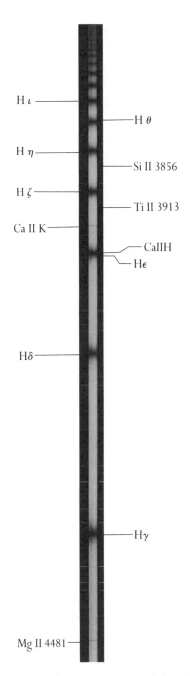

A hydrogen atom's lone electron revolves around its single proton (+) and has an infinite number of possible orbits, of which six are shown here. Each is tightly restricted to a specific energy, shown in the simple graph at the upper right. An electron is mobile between orbits and can be knocked up or down by collision, can jump upward by absorption of a photon of exactly the right energy (equal to the energy difference between the orbits), or jump down with the emission of a photon of that same energy (or wavelength). Upward jumps produce absorption spectra; downward jumps, emission spectra. Transitions that land on or arise from the Balmer series (second orbit), observable in the optical spectrum, are referred to as Hα (6563 Å), Hβ (4861 Å), etc. The Lyman series is found in the ultraviolet; the Paschen and higher-order series are low-energy infrared and even radio.

The spectrum of the star Vega, spectral class A0, for which the largest number of electrons are in the second orbit and the hydrogen lines are at their maximum strengths.

considerable variety. TiO dominates the M-star spectra, and we find CH and CN, among others, in the solar spectrum.

Once astronomers and physicists understood the processes that produce stellar spectra and were able to formulate the rules that control excitation and ionization, they could use the strengths of the absorption lines to calculate the physical conditions within the stellar gases and the all-important chemical compositions. The scientists found a pleasing uniformity of nature. Of the 92 natural elements, 74 are observed in stars. The rest are certainly there, but their abundances are too low to render their lines visible. In gross terms, the majority of stars have the same compositions,

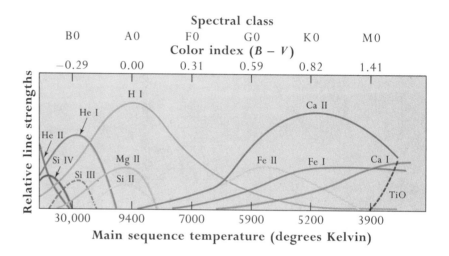

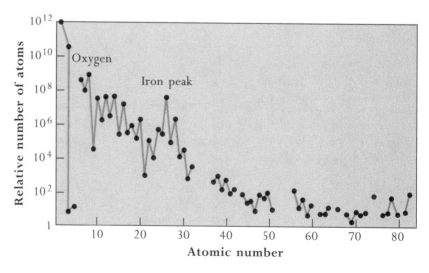

Top graph: *Different atoms, ions, and molecules produce their spectra at different temperatures. Molecules are seen strongly only in class M. At low temperatures we see neutral metals, but hotter stars create more highly ionized species.* Bottom graph: *The abundances of the elements in stars decrease more or less steadily with increasing atomic number except for a great deficiency between helium and carbon and a sharp rise around iron.*

hydrogen dominating at 90 percent by number of atoms, followed by helium, the next simplest atom, at 10 percent. All the other chemical elements are found in the remaining 0.1 percent or so. (The bizarre nature of the Earth, with little H and He, can now be appreciated.) Generally, the more complex an atom, the less there is of it. Oxygen is the next after helium at roughly 5×10^{-4} times the abundance of hydrogen, followed by carbon, neon, and nitrogen. The heavier metals are down by one or more factors of ten, with iron dominant.

The deep red stars are different, however, providing a portent of fascinating theoretical discoveries to come. The N stars, not a part of the standard spectral sequence, differ from class M not by temperature but by *composition*. Their carbon-to-oxygen ratios are reversed, the result of interior stellar processes. The R stars are warmer carbon-rich versions and correspond in temperature to classes K and G. In the 1930s yet another odd kind of star was discovered, labeled class S, which is intermediate in carbon composition between M and N, and at the same time has its TiO replaced by zirconium oxide, ZrO. All the standard spectral classes are now in place.

Stellar spectra provide the promised method of measuring radial velocities. The precise wavelengths of tens of thousands of the absorption lines of the observed elements and molecules have been measured in the laboratory. If the star is moving along the line of sight, the wavelengths of the absorptions will be shifted by the Doppler effect. The shifts can be measured with a calibration device in the spectrograph that produces emission lines of gaseous iron or another element (these can frequently be seen in the illustrations flanking stellar spectra). The radial velocity can then be determined by applying the Doppler formula. We typically find shifts of a fraction of an Ångstrom and speeds again in the neighborhood of a few tens of kilometers per second, consistent with the transverse velocities.

THE HR DIAGRAM

Since the luminosity of a radiating body depends so strongly on temperature, we would expect to see a clear correlation between absolute magnitude and spectral class—and we do. But there is a delightful surprise. The first investigations were made early in this century by the Danish astronomer Ejnar Hertzsprung and independently by Henry Norris Russell, the dean of twentieth-century American astronomy. In 1913, Russell plotted absolute visual magnitude against spectral class for stars of known distance. His graph shows the anticipated effect, the stars tending to lie along a broad

Henry Norris Russell (1877–1970).

band that climbs up and to the left, visual magnitude dropping (and luminosity increasing) with increasing temperature.

But look! There is another band that climbs up and to the *right*, the brightness increasing with *decreasing* temperature. The only way that can

Orange Aldebaran, the gleaming eye of Taurus, lies in front of the stars of the Hyades cluster that make the bull's V-shaped head. Aldebaran is a common K5 giant almost 50 times the dimension of the Sun; it would fill the solar system to nearly the orbit of Mercury. The Pleiades are at upper right.

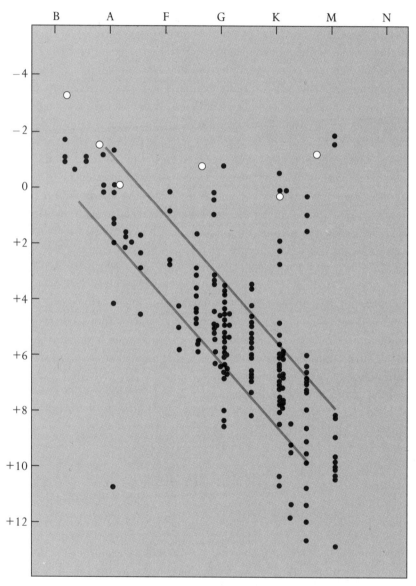

Henry Norris Russell's original diagram, in which he plotted absolute visual magnitudes of stars against their spectral classes. The dwarf sequence (now called the main sequence), which includes the Sun, goes from lower right to upper left. The giant branch scatters out toward the upper right. One white dwarf sits at the bottom.

happen along this locus is if the stars become larger as temperature drops to offset the effect of the rapidly decreasing stellar flux. Russell, following Hertzsprung's earlier suggestion, discriminated between these two bands by calling their stars dwarfs and giants, thereby ensuring a wonderful stellar nomenclature. The Sun occupies the main band, and in spite of its tremendous size is still a G2 dwarf. The giants, however, are of *truly* impressive dimensions. Their diameters can easily be estimated from the blackbody equation, $L_\star = 4\pi R_\star^2 \sigma T_\star^4$, where $\star$ symbolizes a star. If we place everything in solar ($\odot$) units, the constants disappear, and $R_\star/R_\odot$ is just $\sqrt{(L_\star/L_\odot)}/(T_\star/5780)^2$. The effective temperature comes from the spectral type (or the color), $L_\star$ from M_{bol}, and we instantly have the radius. Look at the sky's archetypal giants, orange Arcturus (K1) and Aldebaran (K5). With an absolute bolometric magnitude of -0.3, Arcturus is 105 times more luminous than the Sun. Its great brightness and its relatively cool temperature of 4400 K yield a radius 18 times solar; and Aldebaran is more than twice as big as that. Mira, a cool M7 giant, is *300* times the solar size and would roughly fill the orbit of Mars. Only the giant branch contains the carbon-rich N and S stars, important evidence to be applied later to the mystery of stellar life and death.

Some 15 years earlier, Antonia Maury, one of Pickering's group at Harvard, had noted that some of the hotter stars had broad hydrogen lines, whereas those of others were quite narrow. In 1907, Hertzsprung found that narrow-line stars have considerably smaller proper motions than the broad-line stars, implying that they are much farther away and brighter. These rare and amazing bodies turned out to be ten to a hundred times more luminous and up to ten times larger than even the giants. These supergiants, if plotted, would sprinkle across the top of Russell's graph, which long ago became known as the Hertzsprung–Russell or HR diagram.

The red supergiants are the largest known stars. Apply the blackbody equation to Betelgeuse, the star that marks Orion's right shoulder. Visually, it is 11.8 magnitudes, 54,000 times brighter than the Sun; yet at 3300 K each square meter produces only $\frac{1}{10}$ the solar flux. Bolometrically it is another 1.4 magnitudes brighter, and the blackbody equation yields a star with a radius of almost 7 AU—over 1300 times solar, over four times the dimension of Mira, large enough to overfill the orbit of Jupiter by 30 percent. And that is not the limit! A fourth-magnitude red jewel in Cepheus, μ Cephei—the "garnet star"—shines 1.2 absolute magnitudes brighter than Betelgeuse and is 1.7 times larger, 11.8 AU in radius, comparable to the orbit of Saturn. The volume of such a behemoth could contain over *one billions* Suns. The angular diameters of supergiants can be several hundredths of a second of arc, easily measured by interferometry.

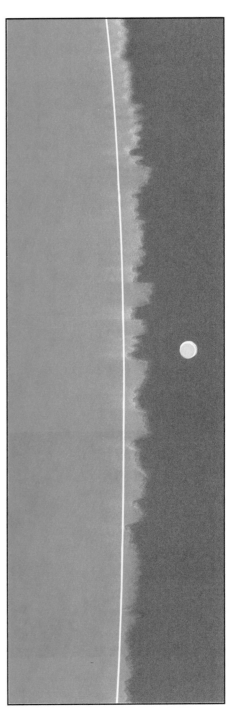

The Sun compared to the surface of Mu Cephei, which on this scale is some 8 meters across.

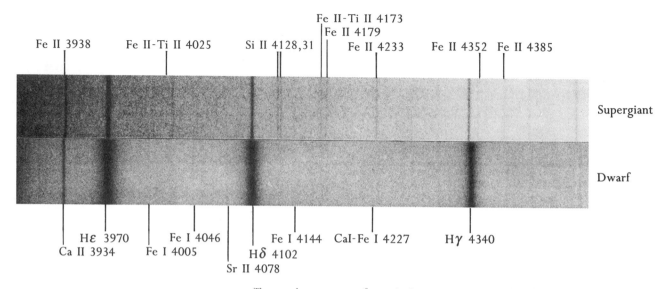

Fe II 3938 Fe II-Ti II 4025 Si II 4128,31 Fe II-Ti II 4173 Fe II 4179 Fe II 4233 Fe II 4352 Fe II 4385

Supergiant

Dwarf

Ca II 3934 Hε 3970 Fe I 4005 Fe I 4046 Sr II 4078 Hδ 4102 Fe I 4144 CaI-Fe I 4227 Hγ 4340

The most luminous stars of type A, the supergiants (HR 1040, above) have much narrower hydrogen lines than do the dwarfs (θ Vir, below) a response to size and density.

The M supergiant Betelgeuse appears here in what may be the first true image of the disk of another star, mathematically reconstructed from observations made with a sophisticated interferometer. The angular diameter can be measured easily.

Return now to the dwarf band, which has since become known as the main sequence. Along it, the stars brighten more than would be expected from temperature alone, indicating that their radii increase as well. Vega, for example, is 2.5 times the size of the Sun. At the top of the sequence we find HD 93129A, $V = -7$ and $M_{bol} = -11.3$, 2.6 million times the solar luminosity. Only part of this extraordinary brightness derives from the star's extreme temperature of 48,000 K. The rest is the result of a dimension 25 times solar, comparable with the smaller giants (yet because of its placement on the HR diagram HD 93129A is still termed a dwarf).

At the other end of the main sequence are dwarfs that live up to their names. Proxima Centauri's reddish color and complex molecular spectrum indicate an effective temperature of some 3300 K. The bolometric luminosity is $\frac{1}{1800}$ that of the Sun, giving a radius only 0.07 solar. Then drop down near the end of the main sequence, to LHS 2924, which has a temperature of only 2600 K and a bolometric luminosity $\frac{1}{36,000}$ solar; its radius is down another factor of 2.5, making it a mere three times the size of Earth. At the distance of the Sun, Proxima Centauri would be just barely discernible as a disk, and LHS 2924 would appear merely as a brilliant star.

Now for something even odder. Down at the bottom of Russell's original diagram is one lonely white star called 40 Eridani B. It is both hot and faint, and as the exact opposite of the giants, it must be very small. This

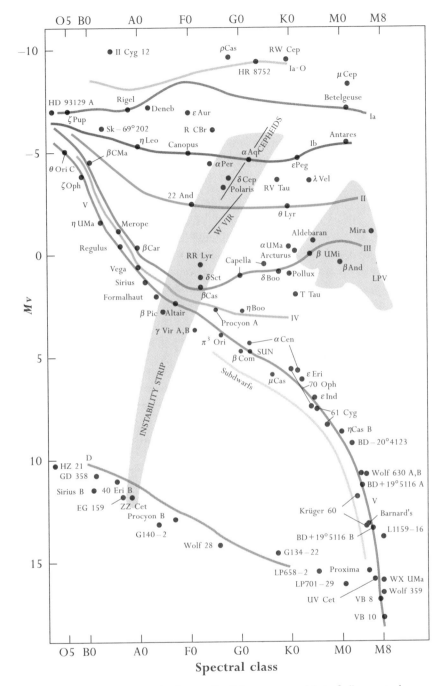

A modern HR diagram shows distribution of well-known stars and loci of all recognized branches. In addition to the main sequence (V) and giant branch (III), we see supergiants across the top (Ia and Ib); extreme supergiants or hypergiants (Ia-0) above them; subgiants between the giants and dwarfs (IV); subdwarfs; and a whole sequence of white dwarfs (D). (Along the main sequence, over two-thirds of stars actually fall into class M.)

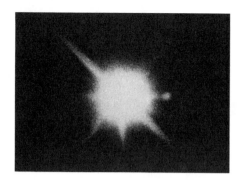

Sirius, the sky's apparently brightest star, has a white dwarf companion 10 magnitudes fainter.

is the first white dwarf, a star roughly the size of Earth. A close look at brilliant Sirius of the northern winter sky, magnitude −1.5, reveals another. This star has a dim companion called Sirius B that is 10 magnitudes fainter, about 2.5 times as hot, and only three-quarters the size of the Earth. As more white dwarfs were discovered, they were seen to string out along the bottom of the graph into the realm of the blue and even the red stars. The name "white dwarf" stuck, though, in spite of the variety of colors.

These various categories of stars define luminosity classes, indicated by a set of roman numerals. Main sequence stars are V, the giants III, and the supergiants I (divided by brightness into Ia and Ib). In between the supergiants and the giants are the bright giants, class II, and between the giants and the dwarfs are the subgiants, number IV. Then to the left of the lower main sequence on the HR diagram is an odd set of stars called subdwarfs. These are low-metal versions of the ordinary dwarfs, their metal deficiency making them appear a bit bluer than normal. The white dwarfs are simply referred to as D. Note that an upper main sequence dwarf can be just as bright as a supergiant. The terms and roman numerals refer to the gross characteristics of the *classes,* not to those of the individual stars. This two-dimensional (temperature and luminosity) classification scheme was

THE LUMINOSITY CLASSES

CLASS	TYPE OF STARS	EXAMPLES
0	Extreme, luminous supergiants; hypergiants	ρ Cas[1]; S Dor
Ia	Luminous supergiants	Betelgeuse, Deneb
Ib	Less luminous supergiants	Antares, Canopus
II	Bright giants	Polaris, θ Lyrae
III	Giants	Aldebaran, Arcturus, Capella
IV	Subgiants	Procyon
V	Main sequence (dwarfs)	Sun, α Cen, Sirius, Vega, 61 Cyg
sd	Subdwarfs	. . .[2]
D	White dwarfs	Sirius B, Procyon B, 40 Eri B

[1] 0-Ia

[2] All faint and obscure

developed by W. W. Morgan, P. C. Keenan, and E. Kellman in the 1940s, and is referred to as the MKK or MK system. The Sun is a G2 V star. At last its classification is complete.

INTO THE GALAXY

The HR diagram provides a powerful means of determining stellar distances. We know the luminosity class and consequently the absolute magnitude of an A star merely by looking at its line width. Measurement of apparent magnitude and application of the magnitude equation then tells us how far away it is. Because of their extreme sizes, the supergiants have much lower atmospheric densities than do the dwarfs. If atoms are close together, they disturb one another, slightly broadening the allowed radii of the electron orbits and smearing the absorption lines. As a result, the dwarfs have broader lines. Each spectral class has such identifying criteria: For example, in class K the giants and supergiants have stronger cyanogen (the CN molecule) absorptions. These standards are all laid out nicely for us by Morgan, Keenan, and Kellman, allowing us simply to look at the spectrum of a star to find its gross luminosity class and thus obtain a spectroscopic distance. We are now in a position to work our way across the Galaxy, and even into other galaxies—providing that we can observe stellar spectra. The distances of all stars listed in Appendix 3 that are greater than about 400 parsecs were obtained in this way.

Parallaxes unfortunately still have a limited range. Within the grasp of this distance method there are no luminous supergiants or O stars, and few B stars, so how do we know their absolute magnitudes in the first place? How do we calibrate the entire HR diagram?

We make use of the profound property of stars to group together into clusters, which abound throughout space. Some are among the most loved objects in the sky: the Pleiades, or seven sisters, in Taurus up and to the right of Orion; the Hyades, which form Taurus' head and surround Aldebaran; the Jewel Box (κ Crucis), next to the Southern Cross.

The Hyades cluster is very close to us, so near that we can not only measure accurate proper motions, we can actually see the convergence of the proper motion vectors, telling us where in space the cluster is moving. Once we know the angle between where the cluster is now and where it is going to be after an arbitrarily distant time in the future, we are able to calculate the cluster's gross tangential velocity from its observed radial

Up and to the right of Orion are the bright B main sequence stars of the Pleiades cluster.

velocity. Combination of the tangential velocity and the average proper motion yields a distance of 47 pc (consistent with the parallaxes of its stars), accurate to better than 8 percent, allowing us to construct the cluster's HR diagram. We can join this value with the Hyades' parallax and add the parallax stars that are not in clusters, the combination providing a powerful fundamental reference.

Unfortunately, the Hyades has no really bright stars, nor any supergiants or even bright giants. But the Pleiades, with a bevy of bright B dwarfs, *does*. To find its distance, we construct an HR diagram with *apparent* magnitudes instead of absolute. We then place the diagram of the Pleiades

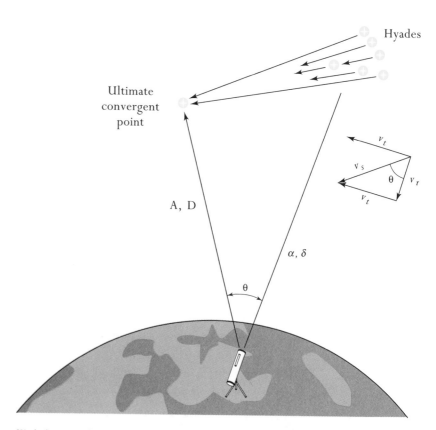

We look out to the Hyades at right ascension and declination (α, δ). The proper motion vectors of the individual stars converge to a point in space with coordinates (A, D) that define the angle between the line of sight and the direction in which the stars are actually going; this is also the angle between the space motion vector and the radial velocity vector (see the diagram to the right). We can then calculate the space velocity (v_s) and the tangential velocity (v_t) from the radial velocity (v_r). The distance of 45 pc follows.

atop that of the Hyades, aligning the lower axes. Magnitudes are logarithmic, so we can slide the Pleiades diagram vertically until the two main sequences fit, allowing us to determine the difference $m - M$, and find the distance by using the magnitude equation. The Pleiades, at a distance of 150 pc, is currently within parallax range, and combination of the two methods and the distance to the Hyades will produce a superb reference base. The next step is to pick clusters that have O stars and supergiants, apply the same method, and there it is: a complete HR diagram.

Once we are able to measure distances (and there are other methods in addition to that based on spectra), we are able to look at the numbers of different kinds of stars. As we climb the main sequence, the so-called luminosity function (the number of stars per unit volume of space per absolute magnitude division) goes down dramatically. It is more meaningful to express it here in terms of spectral class. Over 70 percent of the main sequence stars fall into class M, with masses under one-half solar, and only 9 percent are class G like the Sun. The most luminous stars are remarkably rare: A mere 0.1 percent are class B, and—a stunning statistic—only 0.00004 percent fall into class O! We see a similar phenomenon among the stars off the main sequence. Dim white dwarfs are quite common, whereas the giants are rare compared with the number of stars on the main sequence, and the supergiants are exceedingly few and far between.

These counts are all out of proportion to our naked-eye experience; we see a sky dominated by giants and main sequence stars of classes B and A. This apparent anomaly is a fine example of observational selection, a harrowing problem in astronomy. The dim red dwarfs are everywhere, swarming around us like invisible fireflies, none at all visible to the naked eye. Though the intrinsically bright stars are fairly rare, we can see them for great distances, and so they seem to overpopulate the sky.

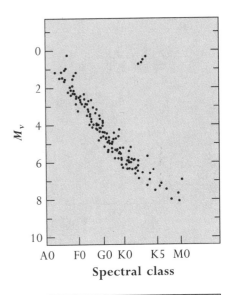

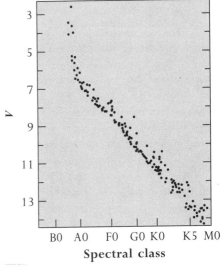

The distances to clusters of stars can be found by the technique of main sequence fitting. The HR diagram of the Hyades (top), with absolute magnitudes, is established by the moving-cluster method and its parallax. We then plot the HR diagram of the Pleiades (bottom), with apparent magnitudes. If we fit the Pleiades diagram on top of that of the Hyades, we can find the difference $m - M$, and from the magnitude equation, the distance.

MASS

Whether for people, rocks, or stars, no distinguishing characteristic is quite so significant as mass. The mass of a star controls nearly everything, from its luminosity to its surface temperature to the length of its life and the way it will die. How can we determine the masses of bodies that we cannot touch and weigh? Again, begin with the Earth. You certainly cannot put it on a scale. All you have to do, however, is to drop a ball and measure its acceleration in the Earth's gravitational field. Since $g = GM_{Earth}/R^2$, and R and G are known, the mass of the Earth is easily found to be 6×10^{24} kg.

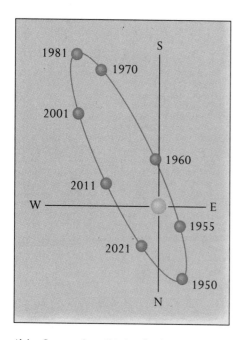

Alpha Centauri B, a K1 dwarf, orbits α Cen A (G2 V); although the orbit is an ellipse, α Cen A is not at a focus. This apparent violation of Kepler's first law results from the orbit's tilt to the line of sight. From the position of α Cen A relative to the focus, the orbit can be rectified, allowing us to find the true path and the sum of the masses from Kepler's third law as generalized by Newton. In fact, both stars have elliptical orbits about a common center of mass.

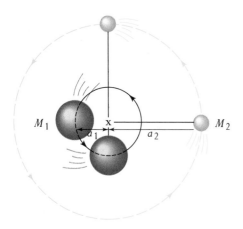

A and G dwarfs orbit one another: $M_1/M_2 = a_2/a_1$.

Simple enough, but how do you apply this concept to the Sun? Again, just drop something, in this case the Earth itself. Remember that an orbiting body is no more than a falling body. To derive the solar mass, use Kepler's third law as generalized by Newton, which requires the Earth's orbital period and semimajor axis. The result, 2×10^{30} kg, is the sum of the masses of the Earth and Sun, in which the Earth's is inconsequential.

This formula is general enough to apply anywhere, to Jupiter's satellites (from which we find that the giant planet is 318 times the mass of the Earth) and to double (or binary) stars. Remember 61 Cygni, α Centauri, Sirius and its companion? They are not rare. At least half of all stars are doubles, triples, or higher-order multiples. If a pair is close enough, we can watch the stars move and map out the orbit. The Alpha Centauri pair completes a circuit in 81 years, Sirius in 50, others in far fewer. P is easy to find, and a is known if we have the distance. Then follows the sum of the masses.

It is simpler to think of the more massive of two bodies as being at the focus of the orbit. In fact, however, both bodies must revolve about each other, with a point in between them—the center of mass—at a common orbital focus. The position of the center of mass depends on the ratio of the masses. For example, the Moon can be said to have an orbit about the Earth with a semimajor axis a equal to 384,000 km. But the Moon's accelerating mass is $\frac{1}{81}$ that of the Earth, so the center of mass is $\frac{1}{81}$ of the way from the terrestrial center to the lunar center. The Earth's orbital semimajor axis is then 4600 km and that of the Moon 379,000 km. If through careful measurements we can locate the center of mass of a double star, we will be able to determine the mass ratio, which, when combined with the mass sum from Kepler's third law, yields the individual values.

A large number of binaries are unresolved: that is, they are so near one another that we cannot see the separation between the components. If they are very close they must move very quickly in order to maintain orbit. The variations in their radial velocities are then observable, and though we see only one star, we might see two sets of stellar spectra that Doppler-shift back and forth as the stars swing around each other, changing their directions. If we are looking at the orbit edge-on, the radial velocities give the orbital velocities, which with the period give the semimajor axes and the masses. Unfortunately, the orbital orientation is unknown, but at least we can get mass ratios and some statistically valid information.

The real treat comes when an orbital plane is in the line of sight. Then the two stars can get in the way of each other and produce eclipses that are sometimes very dramatic as the apparently single star dims by one or even several magnitudes. The celestial hero Perseus slays the monster by showing

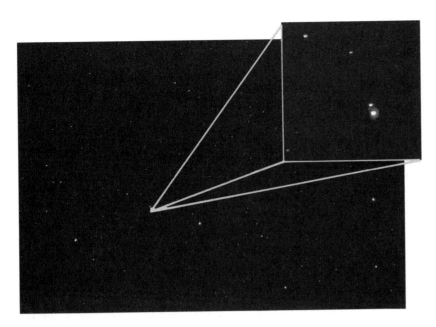

The Big Dipper glides across the open shutter of the University of Arizona's 2.3-meter telescope on Kitt Peak. Mizar, the second star from the end of the Dipper's handle, has a naked-eye binary companion called Alcor. Inset: A small telescope shows Mizar itself to be double, the two members of the system rather widely separated by 14 seconds of arc.

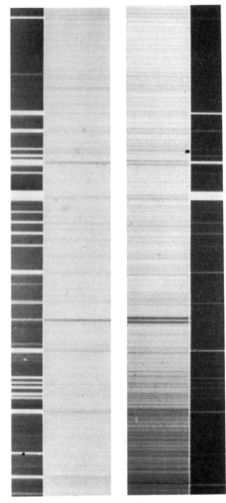

Two spectra of the brighter star in Mizar show it to be a spectroscopic binary. On the left, the lines are single, the two orbiting stars moving across the line of sight. On the right, the stars are both moving along the line of sight, one toward us, the other away. The lines from each star are then Doppler-shifted in opposite directions. Mizar's other component is also a spectroscopic binary.

him the gorgon's head, represented in the constellation by β Persei, Algol, *al ghoul,* the "demon star." Every 2.9 days it drops from second magnitude to third as a small but bright star is partially eclipsed by a huge companion. There are enormous numbers of eclipsers in the sky, and they are a treasury of information. Since we know the orbital orientations from the eclipses, we can get actual velocities from the Doppler shifts, orbital sizes, and stellar masses. From the durations of various phases of the eclipses and the known orbital velocities, we can also get another measure of the physical diameters of the stars. Some eclipses last but a few hours, showing that the stars are small. VV Cephei, just visible to the naked eye, is in eclipse for over a year, revealing a star comparable in size with the monster Mu Cephei.

After all the labor, the thousands upon thousands of measurements, we find a major key to stellar astrophysics, a strict relation between luminosity and mass for main sequence stars. Bolometric luminosity climbs roughly as the mass raised to about the 3.5 power (the exponent is actually a function of mass). A star like Sirius A, 2.3 times the solar mass, is some 20 times as bright! The main sequence is really a *mass* sequence.

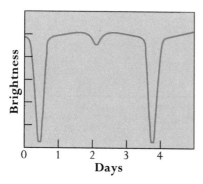

One of the two orbiting components of the star U Sagittae is large, the other small and bright. The latter can hide behind its large companion, producing a primary eclipse every 3.3 days. When the small one passes in front of the large star it occults only a portion of its surface, leading to a much smaller secondary eclipse halfway between the primary eclipses. The shape of this light curve provides orbital parameters that allow masses to be deduced from radial velocities. The duration of the primary eclipse yields the size of the big component, and the time it takes for the small star to go into eclipse yields its diameter.

The Sun is in the middle of the mass range. The observed masses run from a minimum of about 8 percent that of the Sun (the smallest that can actually make a self-sustaining star) to several dozen solar masses. The top of the range is very difficult to define. The rarity of these stars ensures statistically that they are all very far away, and binary orbits cannot easily be studied. But by extending the mass–luminosity relation with a little help from theory, we estimate that at the top, near main sequence spectral class O3, stars can reach 120 times the solar mass, allowing them to blaze forth at bolometric magnitudes near −12 and to be easily visible across intergalactic distances. Low-mass stars are the preferred variety and are everywhere. At the other extreme, there may be only one or two stars near the 120-solar-mass limit in our whole Galaxy of 200 billion. Nature simply does not *like* to make massive stars.

The giants and the supergiants also show something of a mass–luminosity relation, although not so well defined. The giants have masses typical of the middle main sequence of perhaps 1 to 6 solar masses, and the supergiants possess masses tens of times greater than that of the Sun. The white dwarfs are perhaps the most astonishing. The mass of Sirius B is easy to find from its 50-year orbit about Sirius A, and the two stars are near enough so that the center of mass is accurately located. Although the dim

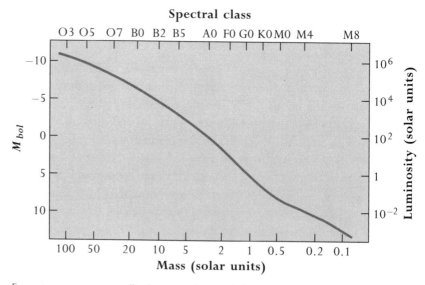

For main sequence stars, stellar luminosity (expressed also as M_{bol} on the left) is a powerful function of mass, as determined from observations of binary stars. On the average, luminosity is proportional to mass raised to the 3.5 power, but the curve is not straight, so the exponent varies considerably. The spectral classes across the top show how mass varies along the main sequence.

white dwarf is smaller than the Earth, *its mass is about equal to that of the Sun.* (This star is heavy; most white dwarfs hover around 0.6 solar.) Its creation would require the Sun to be shrunk into a sphere with less than one-millionth its current volume, thus raising the density by over a million. The Sun's average density is roughly 1 gram per cubic centimeter, about that of water. A cubic centimeter sampled from the interior of Sirius B would weigh a metric ton, 100,000 times that of a similar volume of lead!

VARIABLE STARS

Like people, stars are unique individuals; no two are quite alike. Yet for a large majority we can apply some sort of mean criteria: The G dwarfs, for example, at least have numerous consistent bulk characteristics. But, also like people, there are stars that differ more markedly from the norms, and here and there some really offbeat characters. Deep within the heart of the constellation Cetus, the sea-monster that threatens Andromeda, is a star that winks in and out. For a few weeks out of the year Mira, *o* Ceti, is one of the brightest stars of its constellation, but then it disappears completely from naked-eye view, leaving behind a blank area of sky. This red M giant pulsates with a 330-day period, actually changing its size by about 50 percent and its brightness by some seven magnitudes, a factor in visual luminosity of 600.

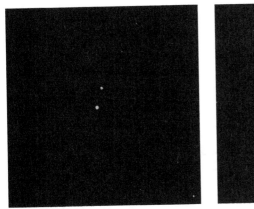

Mira—in Latin, "the amazing"—at minimum (near tenth magnitude) and at maximum (near third). Its variations are obvious even to the naked eye.

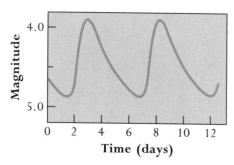

The light curve of δ Cephei, like those of all Cepheids, is remarkably steady. The variation is easy to watch from a backyard. The period of the pulsation tells us the absolute luminosity, from which distance can be found.

It is not alone. Thousands of these long-period or Mira variables are known, all giants of classes M, R, N, or S. Spectroscopy of Mira shows a fairly normal class M spectrum, with the usual TiO bands. Superimposed, however, are bright emission lines of hydrogen. The electrons that belong to hydrogen are jumping down and emitting energy instead of absorbing it. The H atoms are being excited by shock waves (giant sonic booms) generated from below by the pulsation, demonstrating violent mass motions that cause the ejection of gas. The star, in fact, is vigorously losing matter and a goodly fraction of itself into interstellar space.

Next, move just a short distance into the constellation Cepheus, Andromeda's father. In the southeastern corner is third-magnitude δ Cephei, actually a brilliant but distant yellow-white F supergiant. It, too, pulsates and varies, but only by a magnitude every 5.4 days. What is remarkable is the regularity of its variation. Its light curve, the graph of magnitude against time, repeats itself year after year.

Thousands of these Cepheid variables are known and they are among the most important objects in astronomy. In 1912, Henrietta Leavitt of Harvard College Observatory concluded a study of the Cepheids in two companion galaxies of the Milky Way, the Large and Small Magellanic Clouds. These have long been natural astrophysical laboratories; because all the stars within each are all roughly at the same distance from us, we can compare relative stellar properties without knowing the absolute values. Leavitt discovered that Cepheids with longer periods of variation have progressively lower magnitudes, that is, the stars are brighter. This period–luminosity relation can be calibrated with stars in our own Galaxy whose distances we know, or simply through the known distances of the two Clouds (about 52,000 pc for the Large Cloud, 57,000 pc for the Small), which we can find by observing other kinds of stars of known luminosity. The result is a powerful method for obtaining distances. We need only measure the apparent magnitudes (m) and the periods (to get M), and apply the magnitude equation.

The Cepheids are at the high-luminosity end of a great strip of stellar instability that cuts a swath down the middle of the HR diagram. The stars that appear within this strip are unstable against pulsation, and they expand and contract with a concomitant variation in brightness. The pulsation in the instability strip is suppressed by the innate stability of the main sequence, although with care we can see that the stars are still trying to vary. Where the strip crosses the white dwarf sequence, we see the odd ZZ Ceti stars. These tiny bodies chatter away with multiple periods of a few minutes. A Cepheid pulsates radially like a balloon being alternately inflated and deflated. The ZZ Ceti stars are *non*radial pulsators, some parts of the star

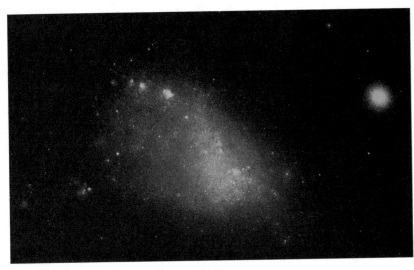

The Small Magellanic Cloud, one of the Milky Way's small companions, lies 57,000 pc away in the southern-hemisphere constellation Tucana. It is close enough for us to resolve individual stars, including numerous Cepheid variables.

expanding, others contracting. We cannot see this bizarre phenomenon, but it is the only explanation for the light curves.

DYNAMICS AND DISTRIBUTION

A look outside at the Milky Way will quickly tell you where most of the stars are. But our Milky Way Galaxy is far more complex than the simple disk introduced earlier. Now that we have the all-important distances, we can begin to pick the Galaxy apart to see where the different kinds of stars are located and how they are moving.

In 1944, Walter Baade, a staff astronomer at Mount Wilson Observatory, was engaged in studying the magnificent Andromeda Galaxy, M 31, with the 100-inch reflector, then the greatest telescope on Earth. Even by then the lights of Los Angeles had largely destroyed the once-superb seeing conditions, but during World War II the city was occasionally blacked out as a defense measure, rendering the skies beautiful once again. Baade had photographed the graceful spiral system through red and blue filters, and to his surprise he found that stars of different colors were differently distributed. Bright blue ones appeared sprinkled throughout the disk and its spiral

The galaxy M 83 displays a bluish Population I within its disk and spiral arms, the Population II central bulge taking on a reddish cast. Sprinkled within the disk are numerous reddish clouds of gas ionized by hot blue stars.

arms, while red ones sparkled in the thick central region. This central portion is seen from photographs of other galaxies to be a great bulge in the disk, looking like a huge, vaguely spherical blister. Different kinds of stars occupy different positions. To discriminate between them, Baade called the blue disk stars Population I and the red ones of the bulge Population II.

We had long seen a similar effect in our own Galaxy. The hot blue O and B stars adhere tightly to the disk of the Milky Way, in Gould's Belt, named for the American astronomer B. A. Gould (1824–1896). Within the disk, they tend strongly to cling together into what are called OB associations that occupy our own spiral arms. Such a grouping can immediately be seen in the constellation Orion, where practically all of the stars are blue. Scorpius is similar. Although the red supergiants are much rarer than the B dwarfs, they lie in the galactic plane as well, and even occupy some of the associations, clearly showing a link between these two apparently disparate groups of stars. With known distances, we can get an accurate picture of the stars' three-dimensional distribution, and we find something quite astounding: The main sequence O and B stars and the supergiants are mostly found in a disk that is only 120 pc or so thick, despite being over *30,000* parsecs across.

Now look at the positions of stars elsewhere on the HR diagram. Other classes define thicker disks. The G stars—those like the Sun—and the K and M dwarfs spread over some 700 pc. The M giants are more extreme and form a disk 2000 pc thick. Moreover, they spread far beyond such confinement into a great spherical halo that completely surrounds the galactic plane and concentrates into a central bulge, giving it a reddish color. We then see M 31 reflected in ourselves and our own Populations I and II. We also see from their galactic distributions that the supergiants are not just luminous extensions of the giants, but are profoundly different kinds of stars.

The same effects are evident if we look only at velocities, those derived so arduously from measurements of radial velocity and proper motion. The O and B dwarfs and the supergiants have low velocities relative to the Earth. They, together with countless dwarfs, including the sun, are all going around the galactic center on roughly circular orbits at hundreds of kilometers per second, and there is little differential among them, a mere 15 or so km/sec. Moreover, the components of their speeds perpendicular to the galactic plane are low, only about 6 km/s, consistent with the thin disk they occupy. But down the main sequence, and in the realm of the giants, the velocity dispersions and the perpendicular speeds are some three times greater, allowing these numerous, cool, reddish stars to spread far away from the plane of the Milky Way and into the halo.

A schematic view of our Galaxy shows a thin disk of bright blue stars (plus a few red supergiants), decidedly clumped together into associations. The white and yellow stars spread farther from the disk, and the red giants extend out into a gigantic halo that coalesces into the central bulge of the disk. Within the disk are thousands of small open clusters of stars, but the halo is dominated by huge, rich globular clusters.

The Population I disk displays a wonderful variety of sights. Not only do we find the loose OB associations, but thousands of much more tightly bound "open" (sometimes called galactic) clusters, those like the Pleiades and the Hyades. A small telescope or even binoculars show that the Milky Way is filled with them. They are sprinkled in among gaseous nebulae, great clouds of gas that are illuminated by the hot O and B stars.

As we concentrate our vision to the central bulge of the Galaxy in Sagittarius, Ophiuchus, and Scorpius, and to the galactic halo—into the realm of Population II—another kind of object takes center stage, the globular cluster. These magnificent stellar gatherings can contain up to half a million stars, packed into a space only two dozen or so parsecs across. They are rare—only about 150 are known in the entire Galaxy—and are generally quite distant. Yet a trio, one in northern Hercules (M 13) and two in the southern hemisphere (ω Centauri and 47 Tucane), are so bright as to be visible even to the naked eye. The telescopic view of ω Cen is extraordinarily beautiful, great masses of stars concentrated so tightly at the center that it is impossible to separate them. Over half the globulars are packed within 60° of the galactic center in Sagittarius. The other half extends an amazing 50,000 pc outward.

The biggest surprise, however, is revealed by the HR diagrams of the globular clusters, which look very different from those of the open clusters or that of the general stellar field of the disk. Most of the main sequence is missing. Instead, there is a prominent horizontal branch that we have not seen before at about absolute magnitude zero that begins at the giant branch and extends leftward toward hotter stars, in many instances as far as class B. All globulars show some sort of horizontal branch, but none of the open clusters does. The two kinds of clusters are entirely distinct from one another; there is no overlap. Some open clusters are so rich that at first glance they may look like poor globulars, but their HR diagrams will quickly reveal them as imposters. They differ spectroscopically, too. A large number (though not all) of the globulars are quite deficient in metals, which

Left: *The Lagoon Nebula, M 8, lies in central Sagittarius. Thousands of these gas clouds, some dozens of parsecs across, occupy the galactic disk.* Right: *Omega Centauri, the greatest of all globular clusters, consists of a million stars in an angular volume only $^1/_2°$ across.*

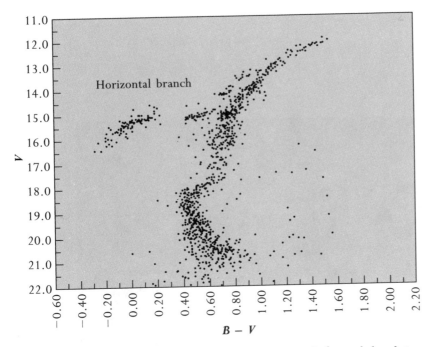

This HR diagram of the globular cluster M 5 uses color index instead of spectral class. It is very different from the diagrams of the open clusters. The main sequence does not go far up, stopping near a color index of about 0.5, which for these Population II stars corresponds to about class G. Most importantly, there is a well-developed horizontal branch that connects to the giants and in this case extends to negative color indices near class B.

relates them to the subdwarfs. The main sequence stars of the metal-poor globulars *are* subdwarfs.

Where the instability strip hits the horizontal branch, it produces a kind of metal-deficient Cepheid called an RR Lyrae star. These vary by up to a magnitude in less than a day. The gap in the horizontal branch of the globular M 5 shows the realm of these pulsators. M 5 has about 100; other globulars have none. All are about the same absolute magnitude and consequently are also excellent distance indicators.

These differences extend to the whole general collections of stars in Populations I and II. Population II by and large has a lower metal content, and its HR diagram has a horizontal branch. It also contains RR Lyrae stars and a form of Cepheid (with its own period–luminosity relation) called a W Virginis star. The lower main sequence comprises subdwarfs, and the upper main sequence is missing. Population I has no horizontal branch, but contains the brilliant massive stars instead. The reason is age and evolution— not just of the stars, but of the Galaxy itself: subjects that we will now begin to explore.

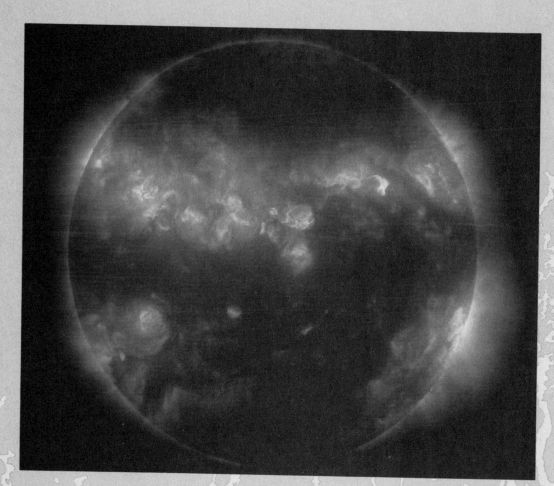

The brilliant solar corona, full of holes and loops, seen in X-radiation.

Chapter Four

TO BUILD A STAR

THE INNER AND OUTER WORKINGS OF STARS, AND STARS AS NUCLEAR FURNACES

*B*ubbling, churning, ringing, alive with energy, the brilliant yellow Sun passes across our sky every day, eliciting little notice: We simply expect it to be there, lighting our day, warming our gardens. We are unaware of the madness going on within and without, oblivious of the intense nuclear furnace that utterly dwarfs all energy resources on Earth, the flaming gases that erupt from its surface, the mighty explosions that hurl great sheets of particles toward our tiny planet. So far, we have seen stars only from the outside, as colorful moving points in a boundless firmament. Now we close in on them, take them apart, and see how they are made, beginning with the most typical of them all, the one that belongs to us.

Look at a summary of its remarkable properties: a diameter of 1.39×10^6 km, 108 times that of Earth; a mass of 2.00×10^{30} kg, enough to balance 332,800 Earths on the scale pans, 745 times the mass of all the planets put together; a surface temperature of 5780 K; and a luminosity of 3.83×10^{26} watts, which relates to an absolute visual magnitude of 4.83, sufficiently bright to be seen with the naked eye at a distance of 20 parsecs.

Perhaps the most astonishing feature is the Sun's durability. It is a relatively simple matter to measure the age of a planet. Most atoms have unstable isotopes that break up into lighter ones with the emission of radiation and are hence called radioactive. Beyond bismuth in the list of elements (atomic number 83), all isotopes are radioactive—some, like radium, dangerously so. All rocks contain some portion, however tiny, of these radioactive elements. Consider the nucleus of the common isotope of uranium, ^{238}U. It will eventually decay, with many intermediate products, into an atom of lead (^{206}Pb) and several helium atoms. Statistically, the decay rate is constant and is specified by the element's half-life, the time it takes for half the atoms of a given quantity to disintegrate: A kilogram of

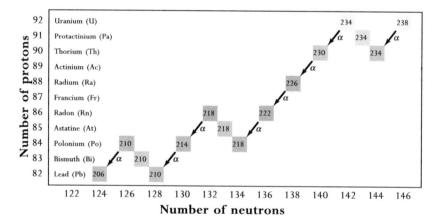

The graph plots atomic number against neutron number for heavy isotopes, showing how the common isotope of uranium, ^{238}U, turns into lead. The ^{238}U first ejects a helium nucleus, also known as an α particle, and turns into ^{234}Th (thorium). One of the neutrons in the thorium isotope then ejects a negatively charged electron. The neutron thus turns into a proton, the atomic number goes up by 1 with no change in atomic weight, and the atom becomes ^{234}Pa (protactinium). Since electrons were originally called "β particles," this reaction is known as β decay. Another β decay turns the substance into ^{234}U. After successive α particle ejections and β decays, the chain arrives at lead.

^{238}U will reduce itself to half a kg in 4.5×10^9 years. When a rock solidifies from its molten state, it seals in the ambient elemental and isotopic abundance ratios. As time passes, the ratio of ^{206}Pb to ^{238}U will steadily increase, a sure measure of the age if we just know the half-life. The oldest rocks on Earth, as determined by several isotopic ratios, are about 3.8 billion years old. The Moon, which is smaller and solidified earlier, gives us 4.3 billion, and the meteorites, rock and iron shards of small smashed bodies from between the orbits of Jupiter and Mars, hold the record at 4.5 billion. The great regularity of the solar system, its planets orbiting in the solar equatorial plane in the same direction as the solar rotation, clearly implies that everything formed nearly at once and that the Sun must also be 4.5 billion years old. The fossil record shows that life formed on Earth somewhere around 3.5 billion years ago, indicating that the Sun has undergone no really gross changes since then and that it has been radiating at a rate of something close to 10^{26} watts *for all that time,* emitting enough energy to run our entire modern world for 10^{20} years.

What is it that gives the Sun and the stars such awesome power? Let us begin to probe inside to see, beginning with the solar surface layers, whose properties will, with suitable application of laboratory data and theory, tell us how the interior functions and what the source of power must be.

The Sun shows a wealth of detail, including a darkened edge, myriad sunspots (some of which are comparable to the Earth in size), bright patches called faculae, and a distinctly textured, granulated surface that demonstrates active convection.

PICKING THE SUN APART: THE SURFACE LAYERS

No attempt should ever be made to look at the Sun without proper filtration; even a short glance can cause permanent blindness. On a misty morning, though, you can sometimes watch it as it rises red above the horizon, the brilliant harsh rays absorbed and scattered by our thick hazy atmosphere. It looks like a solid ball, the edge, or limb, sharp as a razor, an impression reinforced by telescopic examination. Yet, remarkably, the Sun is gaseous throughout, the apparently hard surface an illusion caused by gases of extremely high opacity. The radiation we see as sunlight has worked its way out slowly from the dense interior of the Sun a fraction of a centimeter at a time as it is successively absorbed and re-emitted by atoms along the way. It is finally released from a highly opaque boundary layer called the photosphere—literally, the "sphere of light"—only a few hundred kilometers thick, which at the distance of the Sun corresponds to but a fraction of a second of arc, small enough to be unresolvable. All stars have photospheres, although they are not all so thin as that of the Sun. The

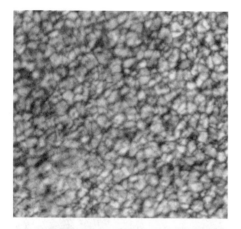

Top: *Solar granulation is the result of rapidly moving convection currents that carry solar energy upward from deeper layers.* Bottom: *A photograph of the Sun taken in the center of the Hα line (a spectroheliogram) records only its emission from the chromosphere, plus the enormous turbulence and millions of tiny spicules that stab into the corona.*

effective temperature of a star is always that of the photosphere, the zone from which the colorful visible starlight originates.

The spectrum gives indisputable evidence about the character of the photosphere, since the absorption lines, which give us its chemical composition, can be formed only in a gas under relatively low pressure. Even a simple white-light (unfiltered) photograph provides a clue. Remember that the Sun is not a solid disk but a gaseous sphere: As we look toward the edge, our line of sight penetrates at an angle to the shallower, cooler layers that radiate at a lower rate. Thus the apparent disk is not uniformly illuminated, but is darkened toward the limb. All stars would show such darkening if they could be seen close up. We can then use blackbody laws to show how temperature climbs with depth. The rate, or temperature gradient (at the surface about 3 K per km), provides data on which to begin to build a mathematical solar model that can be a foundation stone for models of other stars for which such detailed information is not generally available. Since starlight comes from different gaseous strata, each with its own temperature and each obscured by overlying layers in which the opacity changes with wavelength, whole stars cannot actually radiate as blackbodies—hence the need for the contrived "effective temperature."

The Sun is not at all what it seems. On the horizon it looks smooth and perfect, but through the telescope we see great texture, a million tiny bright points called granules at most only a second or two of arc across embedded in a darker matrix. These are the tops of vast convection cells several hundred kilometers wide and deep. They sweep heat upward in hot columns of expanding, rising gas that release their energy by radiation, then cool, darken, and fall. Time-lapse movies show a churning cauldron, each bright cell lasting for but a few minutes before it is replaced by another, and high-resolution spectra display absorption lines that subtly shift in wavelength in response to ever-changing velocities. These granules are grouped into much larger and deeper convection cells called supergranules. And whole areas of granules seethe up and down in large-scale oscillations, the solar surface ringing like a bell in many overtones.

At least twice a year, our orbiting Moon passes between us and the Sun, casting its shadow toward the Earth. By wonderful coincidence, the angular diameters of the Moon and Sun are almost exactly the same, so that under favorable circumstances the shadow can just barely reach us and cut off the light of the photosphere during a total eclipse. As the brilliant surface disappears, the edge of the Moon is suddenly outlined by a narrow deep red ring of light, the "sphere of color," or chromosphere. It is in turn surrounded by a beautiful white solar "crown" or corona that extends outward to many solar radii. Neither of these outer solar layers can be seen

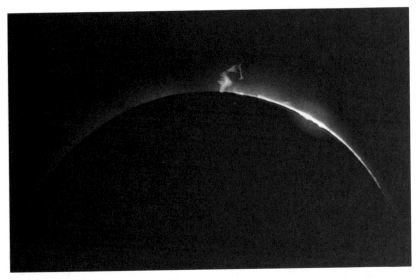

The solar chromosphere, highly reddened from the light of the Hα line, peeks around the limb of the Moon during a total eclipse of the Sun.

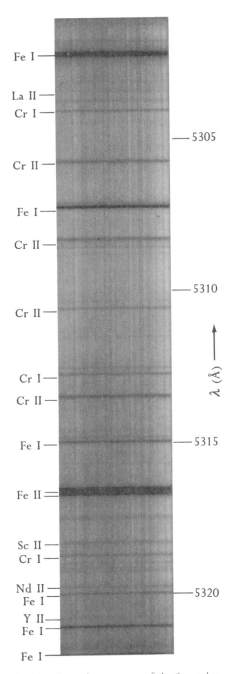

A highly dispersed spectrogram of the Sun, taken with a narrow slit. The vertical striations are caused by granulation. Spectrum lines appear to wiggle (as a result of the Doppler effect) as our line of sight encounters convecting gases.

in ordinary daylight without special instrumentation because their feeble glow is overwhelmed by the light of the blue sky. The corona is a million times fainter than the Sun itself.

The corona and chromosphere defy expectations. Photospheric limb darkening shows that temperature decreases outward. We would naïvely anticipate that as we journey above the photosphere and approach the apparent chill of outer space the rapidly thinning gases above the solar surface would cool. And for a while that is true, the temperature of the upper photosphere dropping to 4200 K. Then, remarkably, the temperature begins to *rise* and, in the chromosphere, stabilizes at about 7000 K over most of its 6000-km thickness. Still farther outward, where the chromosphere makes the transition to the corona, the temperature shoots up to an extraordinary million degrees, reaching an average of 2 million degrees in the corona, with hot spots approaching 5 million.

Recall that flux of radiation depends on the fourth power of the temperature: Why, then, are we not fried here on Earth? In order for a gas to be a blackbody, it must be dense and opaque to its own radiation. The photosphere is so thick that it looks like a solid surface, radiation escaping only from the top. However, the low-density corona and chromosphere are transparent, so blackbody rules do not apply. There is simply not enough gas to produce significant luminosity. The temperature of the corona ap-

The goal of the eclipse watcher is the solar corona, the extensive, low density gaseous halo surrounding the sun. It reaches an astounding temperature of 1 million to 5 million K.

plies only to the energies and velocities of the atoms in the gas, and is properly called a kinetic temperature.

Because of their very low densities and low opacities, neither the chromosphere nor the corona can create absorption lines. Instead of absorbing radiation, these layers *emit* it, and when these layers are presented without the photosphere in the background, we see the photospheric absorptions replaced by bright emission lines. The spectrum of the chromosphere is dominated by hydrogen, in particular Hα at 6563 Å, which gives this transition zone its red color. The element helium was actually discovered in the sun by its chromospheric emissions (the photosphere is too cool to allow helium absorptions), hence its name, from the Greek *helios* for "sun." The corona is so intensely hot that we find emissions of iron and other metals that have been ionized 13 or more times. More remarkably, the high coronal temperature is responsible for copious production of X-rays. Indeed, X-ray imaging from above the Earth's protective atmosphere reveals nothing of the photosphere, but instead shows the whole corona. It is a far more complex structure than can be seen by eye during an eclipse, with vast loops, whorls, and holes. These outer layers are clearly electromagnetic phenomena. The tenuous coronal gases are constrained by great looping magnetic fields that emerge from the photosphere and are heated by the constant release of magnetic energy.

In spite of their tumultuous natures, all these features are collectively referred to as the quiet Sun. They are quiet in contrast with a superimposed set of erratic magnetic phenomena called the active Sun, which we must examine in order to make sense of the chromosphere and corona.

SOLAR ACTIVITY

When Galileo first turned his primitive telescope to the Sun in 1610, he discovered that, contrary to the dogma of the day, the surface was not perfect, but marred with black spots. Several can be seen in the white-light photograph (on p. 107), surrounded by bright areas called faculae (from the Latin for "torches"). The spots tend strongly to come in pairs and to cluster into groups. Individuals range in size from the just barely visible to occasional monsters that are several times the size of the Earth and can be seen with the naked eye. A spot is composed of a depressed dark center, the umbra, surrounded by a grayish halo called the penumbra that slopes upward into the bright photosphere and displays light and dark radial lines. The umbra is far from black, but its temperature of about 4500 K makes it look so in contrast to its brilliant surroundings. A particular spot or spot group will last anywhere from a few days to months depending on its size. The spots are excellent indicators of solar spin, and from them we find that the Sun is a differential rotator, the equatorial regions going around in about 25 days and the period increasing to near 30 days toward the poles.

After Galileo's discovery of the spots, little mention was made of them until the early 1700s, and it was not until 1851 that Heinrich Schwabe discovered that their appearance is cyclic, with a period now known to be

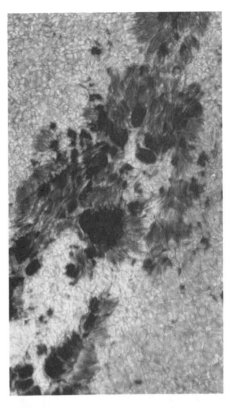

A close-up of a sunspot group shows dark 4500 K umbras, somehow cooled by powerful magnetic field loops stretching from deep in the Sun (field strengths can be up to 5000 times that of Earth). The penumbras exhibit striations radiating outward from the center along magnetic lines of force. Gas can actually be seen streaming away from the umbras along these lines into the photosphere. Convective granulation is everywhere.

The sunspot cycle is quite erratic. The period can vary between 10 and 12 years; some cycles show much greater numbers of spots than others. We are now coming down from the very active 1990 cycle. Between about 1645 and 1715 the cycle seems to have disappeared altogether: the Maunder minimum (after E. Walter Maunder).

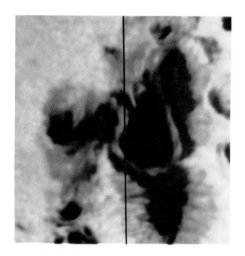

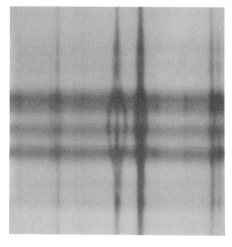

Top: *A large sunspot complex, the dark vertical line showing the placement of a spectrograph slit.* Bottom: *The absorption lines of the resulting spectrum. In the magnetized spot, the Zeeman effect has split the lines into multiple components. Field strengths can be determined from the degree of splitting.*

about 11 years. The spots of a new solar cycle appear far from the equator at about 50° north and south solar latitude. As their numbers increase, their average position creeps toward the equator. At maximum, over a hundred may appear at a time. Finally, at minimum, only a few spots are seen, about 10° or so on each side of the equator, and the next cycle simultaneously begins at high latitudes.

The solution to the mystery of the spots began to be revealed in 1908 by George E. Hale (the man after whom the 200-inch telescope is named), when he found their spectrum lines to be split by the Zeeman effect, which is seen when radiation is created in the presence of a magnetic field. The electrons of an atom act as tiny electric currents, and their orbits and the absorption lines they produce are split by the magnetism. The degree of splitting is proportional to the field strength, and the light within the components is polarized according to the field direction. By examining line-splitting and polarization across the entire Sun, solar astronomers can generate detailed magnetic maps. These show powerful fields whose strengths can range up to 3000 times that of Earth and that loop out of one of the spots of a pair and back into the other. In a way still not understood, the intense magnetism suppresses the upward convection of warm gases and acts as a giant refrigerator. Such a field, sometimes encompassing an area larger than our planet, requires an electrical current flow of over 10^{12} amperes.

The magnetism is highly organized. The Sun has a general, weak dipole magnetic field (one with opposing positive and negative poles) that, per unit area, is only about five times that of Earth. If the north rotation pole of the Sun has a positive magnetic polarity, the leader spot of the pair in that hemisphere—the one that is ahead in the direction of rotation— will always be positive, the follower negative. In the southern hemisphere, the polarities will be reversed. This pattern will hold for the entire 11-year cycle as the spots in each hemisphere approach one another at the equator. Then, for the next 11 years, in the new cycle, *all the polarities will be exchanged,* including that of the general polar magnetic field. The spot cycle is then seen to be a magnetic cycle that is actually 22 years long.

The spots are but one example of solar activity, and for all the energy involved, they represent a relatively benign one. For the real action we look to the corona, as instruments aboard the orbiting *Skylab* satellite did for six months in 1973 and 1974. Several times a day near maximum in the spot cycle, Earthbound instruments detect small bright flashes of light called flares in the chromosphere above and between spot pairs. On rare occasions a flare may grow to terrestrial size and last for a few hours. Solar X-ray images show that the flares originate high in the corona above the active

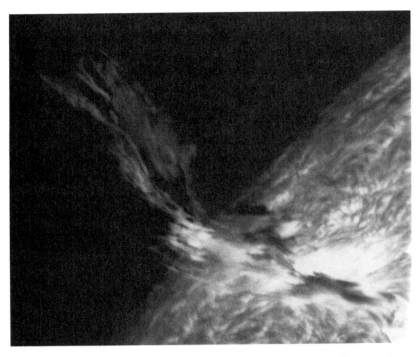

In the chromosphere of the Sun, a powerful solar flare is seen—the result of an intensely hot electromagnetic explosion in the corona producing vast quantities of X-rays that in turn brighten the chromospheric gases.

spotted regions in crowded, tangled magnetic fields that suddenly release their energy in a great electrical spark; in ten minutes the local temperatures can climb to an astounding *20 million degrees*, hotter than the solar center. The event can accelerate electrons to over a third the velocity of light and generate X-rays that slam into the chromosphere, producing the optically visible flares. The chromosphere then boils off some of its gas, supplying material back to the corona, in violence beyond imagination. It is probably this input of magnetic energy and the heat generated by myriad tiny flares—even when the Sun is ''quiet''—that maintains the high general coronal temperature.

Nor do the wonders end there. Associated with the activity are great prominences, cool sheets of gas that condense out of the corona above the active regions. Some are quiet and hang there for days and weeks, while others rain matter down on the photosphere. The collapse of local magnetic fields can also release huge ''coronal mass ejections'' that violently shove blobs of coronal matter from the Sun altogether. The Sun is a true astonishment, constantly changing day to day, minute to minute.

A solar magnetogram, constructed from observations of Zeeman splitting, shows the strengths and polarities (white and black are opposite) of all the magnetically active regions on the Sun. Within a hemisphere, all the polarization patterns are similar; in the opposite hemisphere, they are reversed.

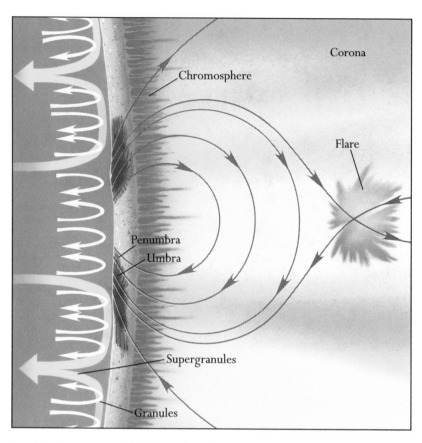

A highly schematic view of the solar surface shows a pair of depressed sunspots created by a magnetic field that loops out of the photosphere into the corona. The powerful magnetism somehow inhibits the photospheric convection. The striations in the spot penumbras lie along the field lines. At the top of the loop, the field lines can connect to produce a flare that lights up the chromosphere below.

In spite of our vast ignorance, we do know that all solar activity is ultimately caused by solar rotation. The Sun is a spinning body made of ionized—that is, electrically charged—gas. The rotation and the churning convection zone, which extends a third of the way to the center, act together, in a way that still eludes us, as a dynamo that produces the magnetic field. At the start of a new cycle, solar magnetism is ordered and relatively simple, with the field locked into the hot gases below the surface. Since the equatorial gases shear past those at higher latitudes, the field lines begin to wind up around the Sun in the direction of the rotation. They are always pushing upward through the solar surface in tiny loops that are intimately related to the granulation, and where they accumulate we see a pair of spots aligned with the equator. The greater the degree of magnetic wrapping, the more the magnetism pops through the surface and the greater the activity. It takes some 11 years for the field to become so chaotic that it breaks down and reorders itself with the opposite polarity, the solar north magnetic pole becoming south and vice versa. Then the whole process starts again, yielding the 22-year observed cycle.

It rather sounds as though we understand something of what is going on, but awareness of our ignorance deepens when we look further into the subtleties of rotation. At any one time there are latitudes on the Sun that are spinning just slightly faster or slower than average, by only a few meters per second. These zones move downward from the polar latitudes, taking 22 years to reach the equator. The spots develop about halfway down in the zones, where the velocities drop. The phenomenon seems to be related to great, longitudinal rolling tubes of gas that extend deeply into the Sun. Why, we simply do not know. But the question, and others like it, is of more than academic interest. All these events touch our Earth and our lives in many ways, some of which we are sure we have not discovered.

For no known reason a prominence will suddenly and explosively erupt into outer space, along with part of the corona.

THE SOLAR WIND

Halley's comet, a periodic visitor to the inner solar system, throws out two beautiful tails, one made of gas streaming directly away from the Sun, the other of dust particles released by the slowly disintegrating nucleus. The comet's gas tail is a creature of the solar wind, which wraps the Sun's magnetic field around it like a tight sleeve and sends it millions of kilometers away from the nucleus.

Once every decade or two, our skies will be graced with one of nature's grandest sights, a bright comet sporting a delicate gossamer tail that can stretch from horizon to zenith. Dozens of faint ones are found every year. These fragile bodies, usually no more than a few tens of kilometers across, are made of dust set within a matrix of ices, mostly water. The comets, which may be the most primitive bodies of the solar system, reside in a pair of huge clouds far beyond the realm of the planets. A large number, however, are on highly elliptical orbits that take them into the inner part of the planetary system and toward the Sun's heat. As they pass the orbit of Jupiter they begin to warm, and the ices sublime into space in massive jets, carrying the solids along with them in a growing cloud that eventually produces a tail that can grow to tens of millions of kilometers in length.

It is not hard to notice that no matter what the direction of the comet's motion, the tail is always pointed away from the Sun. As early as 1950 it was noted that comet tails were likely interacting with a gas streaming from the Sun that constantly shoves the tail radially outward. Theoretical analysis a few years later demanded that the hot corona *had* to be losing matter into space. Confirmation came late in the decade with the detection by satellites of a solar wind, a flow of protons and electrons that blows vigorously outward all the way past the Earth, where some of it is waylaid by the terrestrial magnetic field, and into the far reaches of the solar system. The Sun is actually *evaporating,* albeit at a paltry rate of 10^{-14} solar masses per year (far worse things will happen to it long before 10^{14} years pass). At the Earth, we typically count about 10 or so protons and electrons per cubic centimeter at a (nonblackbody) temperature of 200,000 K howling past us at a speed of some 500–700 km/sec after a journey of about two days. In a sense we are actually *in* the outer reaches of the solar corona. The flow is complex. The corona is tightly confined by the great magnetic field loops and acts as a barrier to the streaming wind. The gases emanate primarily from the coronal holes seen in X-ray images, and the wind's strength is highly variable and dependent on solar magnetic activity.

The wind also highly distorts the Earth's magnetic field, compressing it in front in a shock wave as the prow of a speeding boat compresses the water in front of it. Within this boundary, particles become magnetically trapped in two zones called the Van Allen radiation belts. Closer to Earth, the interaction between the wind and the field also sets up intense rings of electrical current that surround the magnetic poles. To the rear, the wind sends the field into a magnetotail to a length of over 100 terrestrial radii. All the magnetic planets of the solar system are affected by the solar wind, even

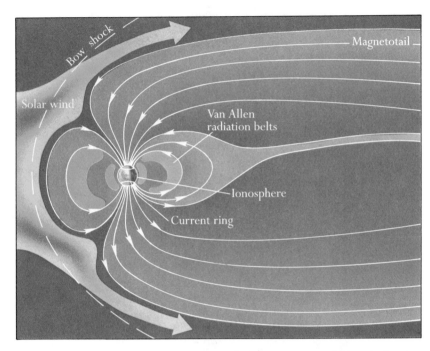

As the solar wind slams into the Earth's magnetic field it creates a powerful bow shock and a long magnetotail, as it does with Jupiter, Saturn, Uranus, and Neptune. The interaction of the solar wind and the terrestrial magnetic field produces the Van Allen radiation belts and drives massive rings of electric current around the magnetic poles.

as far as Neptune, 30 AU from the Sun. The solar influence finally comes to a halt where it slams into the surrounding interstellar gas. We do not yet know where that barrier, the heliopause, lies. Two extraordinary spacecraft, *Voyager 1* and *Voyager 2,* are outbound to find it, perhaps arriving in true interstellar "outer" space about the year 2020.

SOLAR ACTIVITY AND THE EARTH

The night is dark and clear as you stand outside to admire the stars. You look to the north and see an odd red patch in the sky, perhaps the glow from a fire over the hill. But the spot grows and begins to develop streamers that arc over the heavens, the colors changing from red to green to white. Suddenly the night is a theater stage hung with great draperies, lit with pulsing spotlights that illuminate a celestial review beyond all imagina-

The aurora australis, a result of enormous electrical currents that ionize the upper atmosphere, photographed from the flight deck of the spacecraft Discovery.

tion, one that exalts the sky, frames the stars, and testifies to the awesome power of nature—and the Sun. You have just seen a display of the aurora (borealis in the north, australis in the south), the northern or southern lights. An erupting coronal mass ejection sends a shock wave screaming down the flow of the solar wind. About two days later it smashes into the Earth, where it sets off a disturbance that causes magnetic field lines to reconnect close by, depositing a vast amount of new energy into the polar current rings. The result is an electrical display that produces ionization of the upper air. Recombination of the electrons and their parent molecules and atoms then lights the sky. The electrical potentials during such "magnetic storms" are so great that they can cause currents to flow in long power lines that are sufficient to blow circuit breakers and produce power blackouts—making it easier to see the aurora!

Solar–terrestrial relations extend much more deeply. Between 1645 and 1715 the cycle seems to have disappeared. During this period, called the Maunder minimum after the man who first pointed it out, the world was plunged into a "little ice age": Europe was covered with deep snows, and rivers in the southern United States froze. One data point does not make a correlation, of course, but there is additional evidence from the geologic record of radioactive ^{14}C. Its manufacture in the atmosphere by high-energy particles (cosmic rays) from deep space is suppressed by high solar activity, allowing the demonstration that low activity coincides with cooler weather. What will happen when the activity drops again, which it surely will? Do we really want to find out? And why does it happen?

Long-term trends also relate solar activity to droughts, and recently the activity has been great enough to expand the atmosphere sufficiently to bring *Skylab* hurtling to the ground and to necessitate placing the *Hubble Space Telescope* into a higher orbit. We have really only begun to explore this vast subject, so long are the intervals over which we must look and so difficult is the theory. But the effects are profound. To know the Earth and, as we will see, the stars, we must know the Sun.

SOLAR ENERGY

For all of its marvels and mysteries, what we have seen of the Sun so far is only its outer husk. All of its external properties, from its normal radiation to its activity, are driven by what goes on inside. The most important problem is the origin of the solar luminosity. What causes the Sun—and all the other stars—to shine at so great a rate? The first really serious discussion, initiated by the physicists Lord Kelvin and Hermann von Helmholtz in the 1860s, involved the conversion of gravitational potential energy into heat. There is no question that this process must take place. A body the mass of the Sun is held together by its own gravity. The internal squeeze would elevate the outward pressure and raise the temperature to extraordinary values, as high as 40 million degrees. Any body hotter than its surroundings must radiate away part of its internal energy; when it does, it contracts, the shrinkage producing more energy, and the process continues. Only a minute rate of contraction, about 20 meters per year, is needed to produce the Sun's observed power of 4×10^{26} watts. So far so good, but a problem remained: not one of luminosity, but of *time*. Even at this rate, the Sun would have used up all its gravitational energy in only 100 million years. This is an immensely long time and satisfied the requirements of the astronomers and physicists of the late 1800s; but it was soon found to be totally out of line with the age of the Earth. Gravity alone simply cannot do the job.

Understanding had to wait for the early part of the twentieth century, which brought the discovery of the nature of the atom, the development of the theories of relativity and quantum mechanics, and the realization that the Sun and almost all stars are mostly made of hydrogen. The first major

Sir Arthur Eddington (1882–1944).

step was taken in 1926 by Sir Arthur Eddington, the English physicist and astronomer who dominated much of his science in the early part of the twentieth century. He was one of the first to embrace the young Albert Einstein's theory of relativity, which contains the famed formula for the equivalence of matter and energy, $E = mc^2$. The import of this relation is that because of the large value of the speed of light, 3×10^8 m/s, only a tiny amount of mass is needed to create a huge amount of energy, enough to light the Sun. Eddington knew that a helium atom is slightly lighter than four hydrogens—the deficit is 0.7 percent—and he surmised that if hydrogen could be turned into helium, an enormous amount of energy could be created out of the missing mass. The solar luminosity could be produced by the conversion of 6×10^{11} kg of hydrogen per second. While this number seems huge, it is only 3×10^{-19} the mass of the Sun. At that rate, if the Sun were made of pure hydrogen, it could shine for an astounding 10^{11} years, far longer than the age of the Earth.

Such numbers alone, as satisfying as they may be, are still not enough to prove that hydrogen fusion is responsible for solar power. At the time, it was not at all clear that the Sun had enough hydrogen to do the job, and even if it did, no one could say how the transformation actually took place. Cecilia Payne-Gaposchkin solved the first difficulty in the late 1920s with the aid of the new quantum view of the atom, when she showed that hydrogen dominates the stars even though its spectrum lines may be weak or absent.

The second difficulty is more profound. In any gas, the atoms are flying about with an average speed that is given by temperature. At 40 million degrees, the temperature calculated by Eddington under the gravitational theory, the hydrogen nuclei—protons—typically smash into one another at the astounding speed of 2000 km/s. Moreover, there is a great distribution of speeds about the mean, so that a tiny fraction are going 10 or more times faster. But even at the extreme, the repulsive barrier imposed by the protons' similar charges is too great to allow the tiny particles to approach one another closely, let alone stick together. How then can they fuse into a helium nucleus?

This problem was overcome in 1928 by George Gamow, Fritz Houtermans, and Robert Atkinson, working at the University of Göttingen, with the application of quantum mechanics, specifically Heisenberg's famous uncertainty principle. As we have seen, light behaves simultaneously as a wave and as a particle in the form of the massless packet of energy called a photon. Remarkably, subatomic particles have the same kind of duality (with their wavelengths dependent upon the particles' energies). As a result, one cannot simultaneously know both the exact position and velocity of a particle such as a proton or an electron. It is not a matter of imprecision in measuring devices but a fundamental limitation of nature. The uncertainty in location times the uncertainty in momentum (mass times velocity) is of the order of h, Planck's constant. The number is so tiny that the uncertainty cannot be noticed on any large scale, but it has a powerful impact on the atoms.

The probability that a particle is at any given point depends on the amplitude of its wave, which is greatest at the particle's most likely position. However, in principle the wave extends to infinity. Although the amplitude rapidly decreases away from the most favored location, a particle can actually be found anywhere along its wave. By the uncertainty principle, the position of a particle cannot be known: It may be here, it may be there. If two protons can be brought sufficiently close together in a high-temperature gas, one might suddenly find itself somewhere else along its wave and tunnel right through the electrical barrier between them. If they can attain a separation less than about 10^{-13} cm from one another, the strong force takes over, overwhelms the electrical repulsion, and makes the particles adhere to create a new atom or isotope. Tunneling is so efficient that the temperature can be considerably lower than Eddington's calculation: a "mere" 15 million degrees.

However, even the discovery of tunneling still does not reveal how the reactions that produce helium from hydrogen actually proceed and what mechanisms produce the radiation. Several other discoveries were necessary, including that of the neutron, the deuteron (^{2}H, hydrogen with a proton and a neutron as its nucleus), the positron, and the neutrino. The first three of these were isolated in 1932. The positron is a positive electron, the first type of antimatter (matter with reversed charges) ever seen. Oddly, the strangest of the four, the neutrino, was indicated a year earlier. This presumably massless particle, which travels at the speed of light, had been suggested by Wolfgang Pauli to explain a mysterious disappearance of energy in nuclear reactions. Not actually discovered until 1956, it will play important roles in our later assessments of stars.

Eight years after these elementary particles were found, Hans Bethe and Charles Critchfield had the answer to hydrogen fusion (or hydrogen "burning," as it is frequently called) in the proton–proton (p–p) reaction, which involves a stepwise process. The first step is the creation of deuterium by a tunneling collision between two protons. One of the protons instantly turns into a neutron with the ejection of its positive charge in the form of a positron. The process, which is accompanied by the release of some energy in the form of a neutrino, is called inverse beta decay. It is the reverse of beta decay, in which a neutron bound within an atom ejects an

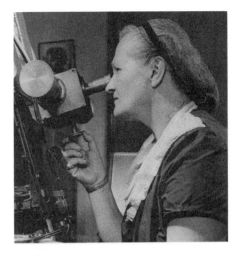

Cecilia Payne-Gaposchkin (1900–1979).

electron (in the old jargon known also as a "beta particle"), thus turning itself into a proton. (Beta decays are in part responsible for the decay of uranium into lead; see the figure on p. 106). Antimatter and normal matter cannot coexist. As soon as the positron encounters a free electron, which it does almost instantly since there are so many available in the intense ionizing heat of the solar core, the two annihilate each other in a flash of energy in the form of a gamma ray.

The wonder of solar power is its stability. Once we begin a reaction in a nuclear bomb, the release of energy is almost instantaneous, but the Sun liberates its energy over billions of years. Why does it not explode too? The key to the puzzle is this first reaction. Even with tunneling, $^1H + {}^1H \rightarrow {}^2H$ is exceedingly slow. A typical proton waits over *10 billion years* before it attains enough energy by random collision to have sufficient speed to overcome the electrical barrier. The process works only because there are so *many* protons in the Sun that at any one time a few are actually making the transition. Once a deuteron is created, however, it reacts with blinding speed. In under a second it encounters and absorbs another proton (with the further production of a gamma ray) that does *not* turn into a neutron. The three particles, two protons and a neutron, are now helium, albeit in a light form, ^3He. In the final step, typically after a million years, two ^3He atoms collide with enough velocity to merge into normal helium, ^4He, in the process ejecting a pair of free protons back into the burning maelstrom. The vast bulk of solar power, about 77 percent, is produced in this way, aided by several other reactions that will have their "place in the sun" later.

This sequence of discoveries, based on the work of a great many people, should still astonish us. By inferring processes involving atoms that we cannot see inside a body that our vision could not penetrate, astronomers and physicists had found the source of the energy that keeps us alive.

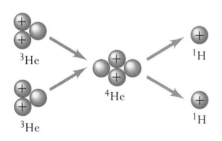

The proton–proton reaction requires the collision of two protons to form a deuteron, ^2H. The deuteron combines instantly with another proton to create ^3He, and two ^3He nuclei eventually combine to form ^4He with the ejection of two of the protons. The net result is one helium atom (four protons) and the release of energy.

THE SOLAR INTERIOR

Let us now build a Sun. We can make a mathematical model and calculate the temperature, pressure, and density at all interior points, so that we can calculate the rate at which hydrogen is converted into helium. The construction of such a model is a demanding procedure that requires the simultaneous solution of several interlocked equations, a task that lends itself admirably to high-speed electronic computers. The basic law is that of hydrostatic equilibrium, which Eddington had used earlier to find the interior solar temperature under the assumption of pure gravitational contraction. In every layer within the Sun, the weight of the overlying gas must be equal to the outward-pushing pressure; otherwise the Sun would either expand or contract, and by simple observation it does neither. One must also adopt an equation of state, a relation among pressure, temperature, and density. Under most conditions encountered in astronomy (though not all, as we will see), the pressure is proportional to the density times temperature, a relation known as the perfect gas law ($P = nkT$, where n is the number of particles per unit volume and k is Boltzmann's constant). The more atoms there are per unit volume and the faster they are going, the greater must be the outward push. Since as we proceed into the Sun, the weight on a layer imposed by gravity gets larger, the density and temperature must correspondingly increase.

Next, we must consider the rate at which energy is generated by hydrogen burning as a function of distance from the solar center; this will depend on the variation of temperature, density, and the compositional fraction of hydrogen fuel. The rate per unit volume will be greatest at the center and will drop as the temperature decreases outward. As we move away from the center and include more and more mass within successively larger shells, the total luminosity generated rapidly increases. Eventually the increase in luminosity slows and comes to a halt when, in our outbound trip, we reach a vaguely defined limit of about 7 million degrees, where the atomic velocities are just too slow to sustain reactions.

Perhaps the most difficult problem involves the way in which energy is transported, by conduction, radiation, or convection. Conduction (direct heat transfer by contact) can immediately be dismissed, since its rate in a gas is very slow. Radiation, however, is vitally important. The rate at which energy flows by radiation depends principally upon the temperature gradient. Place yourself somewhere inside the star. Radiation is coming at you from all directions within the hot, hazy surrounding gases. Temperature always decreases outward, so that as you look in the direction of the surface (you will only see a few centimeters) you view gas that is at a slightly lower temperature than the gas you see if you look inward. Because the flux of radiation depends on temperature, there must be slightly more outbound radiation than there is inbound. As a result, there is always a net flow of energy outward. The steeper the gradient, the greater the flow.

The gradient is controlled by the gas's opacity, which is the degree to which the gas will absorb, or impede, radiation. The higher the opacity, the greater must be the gradient in order to get energy to move outward. The opacity is dependent upon the detailed chemical composition; it can be very difficult to calculate, requiring intimate knowledge of thousands of atomic absorption processes, and is thus one of the limiting factors in building stellar models. We read in the newspapers and science magazines of glamorous discoveries of galaxies, black holes, and the like, but never of the unsung heroes of a massive five-year effort to improve our knowledge of opacities, without which we cannot understand the inner workings of stars.

Eddington pointed out that a flow of radiation must produce a pressure that adds to the gas pressure and plays a role in hydrostatic equilibrium. When an atom absorbs a photon, it also absorbs the photon's momentum (even though a photon is massless, it has momentum because it carries energy). As a consequence, the atom recoils. The radiation flow, which depends on the temperature gradient and the opacity, then provides an outward push. Near the center of the Sun, the flux of radiation is so amazingly high that the radiation pressure is about 10 percent of the gas pressure and helps buoy up the Sun considerably.

All the equations that govern the Sun's physical processes must be solved at the same time. The object is to determine the internal temperature, density, pressure, interior mass, and energy-generation rate at every value of radius, so that given a solar mass and the chemical composition, the calculated solar radius, total luminosity, and effective temperature can match the observed values. We find that the temperature at the center must be about 15 million degrees and that the core of the Sun, the region in which the energy is generated, occupies about 30 percent of the solar radius and about 3 percent of the volume. Because of its astounding density, which reaches 160 grams per cubic centimeter (more than 10 times that of lead!), the core contains 40 percent of the solar mass. In spite of this huge density, because of the enormous temperature *the core is still a gas*. At the very center, about half of the hydrogen has already been fused into helium.

Beyond this central zone, at a point where the temperature drops below about 7 million degrees, we enter the solar envelope, wherein the remaining 60 percent of the solar mass is spread out over about 70 percent of the radius. The envelope acts much as an insulating blanket that impedes the flow of radiation and maintains the core's pressure and temperature. It also degrades the energies of the outbound photons. Energy created in the core works its way out of the Sun by a random-walk process, aided by the very slow push of the temperature gradient. It is rather like a drunk who walks from Los Angeles to New York by bumping into lampposts and trees along the way, gently propelled by a very slight westerly wind. Because of the enormous opacity of the solar gases, even at the speed of light energy

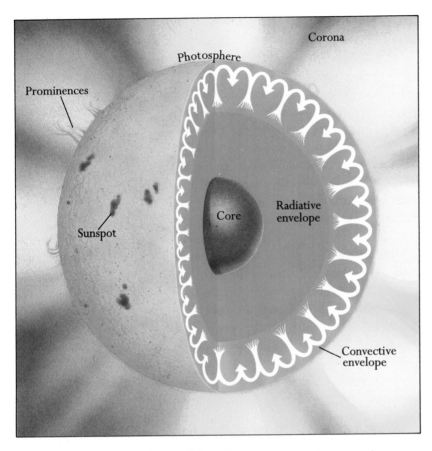

A model of the Sun shows us the size of the nuclear-burning core and its surrounding insulating nonreactive blanket, the outer part of which is in a complex state of convection.

will take over a million years to arrive at the surface! Each time the energy is re-emitted, it comes *on the average* from a layer at a slightly lower temperature. Consequently, because of the blackbody law, the mean photon energy must also be lower. But since the energy cannot be destroyed, the number of photons must go up in compensation. What began as a single deadly gamma ray finally emerges from the photosphere as a thousand low-energy optical photons, the envelope thus converting the radiation of the core—the kind emitted in nuclear bomb explosions—into a spray of soft yellow light.

The complications of the opacities are bad enough, but there is a worse problem. About 70 percent of the way from the center to the photosphere, the temperature gradient becomes such that it is more effective to transport energy not by radiation but by the third possible mode, convection. The gases then begin to roll in a complex series of layers that eventually come to the surface and are seen as granulation and supergranulation. Unfortunately, despite the power of physics and modern computers, we still do not understand well how convection works. We cannot, for example, predict a priori the sizes of the cells, the so-called mixing lengths, and must rely on ad hoc assumptions—those designed to give the "right," that is, the observed, answer. It is the convection layer, combined with differential rotation, that generates the solar magnetic field and solar activity, whose origin we can then hardly understand either. Remarkably, even though convection is confusing, we can probe the layer through analysis of the large-scale solar oscillations and observe its depth directly.

In spite of these problems, astronomers have been able to build a solar model that will both match the observed luminosity and show a lifetime that is consistent with that of the solar system, demonstrating that solar energy is indeed generated largely by the proton–proton cycle. The calculation of the solar age involves a number of considerations. Earlier in this discussion, we assumed that the entire solar mass could be converted into helium and found that the process would take 10^{11} years. However, now we see that only about 40 percent of the solar mass, or $0.4M_\odot$, is initially available, and our estimate of the solar life is accordingly shortened. Fusion proceeds fastest at the center, so the hydrogen there will be exhausted long before the entire core becomes depleted. Finally, we must look at how the p–p fusion rate changes as the available hydrogen diminishes. The pressure of a gas under the perfect gas law depends only upon the *number* of atoms per unit volume, not upon their kind. As four atoms of hydrogen are converted into one of helium, this number decreases, so that to maintain the internal pressure, the core has to contract under the weight of the overlying layers. As a result, the interior temperature increases, so the fusion rate and the luminosity are maintained even as the fuel supply dwindles. It is this process that gives the Sun its stability and that has allowed the time for life to develop on Earth; it is responsible for our very being.

The calculations show that the total lifetime of the fuel in the central core is about 10 billion years. When it is gone and the solar hydrogen-burning stage is effectively ended, the Sun will blossom into a brilliant red giant, a matter to be explored in the next chapter. With the age of the Earth and Sun known to be about 5 billion years, we see that the Sun is distinctly middle-aged, with a long and presumably good life ahead of it. Nature has produced a wonderfully balanced fusion machine, with the unimaginable violence in the core so tightly controlled that we will be able to go outside and enjoy a sunny day for a long time to come.

NEUTRINOS

All our standard observations of the Sun are of its surface or of its surrounding halos, the chromosphere and corona. Even probing the convection layer by the analysis of solar oscillations takes us only partway into the Sun; it provides no information on the core and gives no direct proof that hydrogen fusion really runs the Sun. So far, all we have is the indirect evidence of steady solar luminosity over a very long period of time.

Neutrinos may provide such proof. These are the most noninteractive of all subatomic particles. It would typically take *a light-year of lead* to stop one. Consequently, they are effectively unimpeded by the overlying solar layers and come flying out immediately upon creation. A neutrino telescope would allow us to look directly into the solar heart. But if matter is transparent to the little pieces of energy, how do we detect them?

Matter is not *perfectly* transparent. Neutrinos can induce a variety of nuclear reactions, and even though these are all intrinsically very unlikely, there are so many neutrinos emanating from the Sun (the particle flux at the Earth is an astounding 10^{10} per square centimeter) that we should occasionally see one. The first detection experiment has been running for nearly 20 years, deep in the Homestake gold mine in South Dakota, where Ray Davis set up a tank containing 100,000 gallons of the cleaning fluid perchloroethylene (C_2Cl_4). About a quarter of the chlorine is the ^{37}Cl isotope, which has a particular propensity to capture a neutrino and turn into radioactive ^{37}Ar. Radioactive atoms are detectable when they decay. A count of the number of converted atoms along with the known capture probability and decay rate tells how many neutrinos are emerging from the Sun.

The experiment unfortunately does not count the neutrinos that are produced by the p–p process itself, which have energies that are too low to affect the $^{37}Cl \rightarrow {}^{37}Ar$ reaction, but it does count the much more energetic neutrinos generated by four secondary chains. Still, the solar model tells us with considerable accuracy how many of these there are and how they relate to the more prolific main chain. What has been found over two decades of work is quite disconcerting: The experiment has counted only somewhat more than a quarter of the expected number. There are a variety of possible explanations. The experiment may be flawed, the estimate of neutrino capture rates may be wrong, the Sun's core temperature or chemical composition may be different from that expected (and thus the solar model wrong)—or we do not understand neutrinos.

The question is so important, both in terms of the Sun and the basic model of nuclear physics, that several other experiments have been

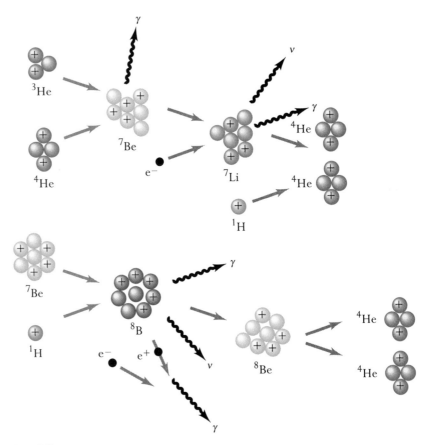

Four different reactions generate the countable neutrinos in the Homestake mine experiment, of which two are shown here. Top: ³He produced in the p–p reaction can combine with ⁴He to yield ⁷Be and a gamma ray. The beryllium is then struck by an electron, which converts one of the protons into a neutron, creating ⁷Li, a gamma ray, and a neutrino. Alternatively, bottom, the beryllium can capture a proton to become unstable ⁸B, which can decay into ⁸Be, a positron, and a very energetic neutrino. Both reactions end in the production of two helium nuclei.

mounted or planned. The next after Homestake, built in 1987 in the Kamioka zinc mine in Japan, uses a large pool of ordinary water. Neutrinos striking the atomic electrons produce a brief flash of radiation that is detected by photoelectric cells lining the walls of the tank. Unfortunately, this detector has an even larger energy threshold than does the Homestake experiment and can capture neutrinos from only two of the side chains. But the results are consistent, in that it finds only about half those expected. It is also directional, and clearly demonstrates that the neutrinos it does detect are in fact coming from the Sun. So the problem does not seem to be with

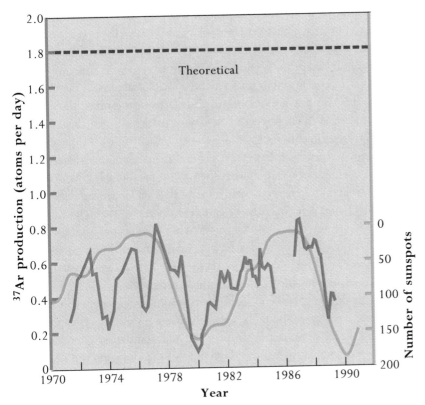

A graph of the number of neutrinos (in terms of the number of ^{37}Ar atoms created per day) over a 20 year period (red) shows the counts to be well below those expected from theory. The variations in the curve seem to match the count of sunspots (yellow) over the solar cycle, the magnetism of the Sun at maximum perhaps suppressing the neutrino count.

the experiments themselves. Further evidence for what is happening derives from long-term plots of the Homestake data. The neutrino flux is apparently anticorrelated with the solar cycle! The Sun may indeed be responsible. But how?

Two experiments, a joint United States and Russian venture in Russia *(SAGE)* and a European experiment in Italy *(Gallex),* use the rare element gallium, and count neutrinos when they convert ^{71}Ga into radioactive germanium, ^{71}Ge. These instruments can detect the neutrinos of the main p–p reaction, and together find about half the expected number. The solar model should not be that wrong. Consequently, we look to the neutrino itself.

Complexity is piled on complexity. We have seen that there are five nuclear reaction chains that generate solar energy (and there is actually one more). Moreover, there is not one kind of neutrino but *three,* and each of them has an antineutrino version that spins in the reverse direction. The kind of neutrino produced in solar reactions is associated with the electron. But there are also heavier versions of the electron, the muon and the tau particle, that have their respective neutrinos, which are not detectable by any of the experiments (neither are their antineutrinos). If the electron neutrinos generated by the p–p cycle have a small amount of mass, then ambient solar electrons can convert them into uncountable muon and tau neutrinos. The more energetic neutrinos of the side chains are less subject to this effect, and consequently we can detect some of them, which would explain why Homestake finds a quarter what is expected and Kamioka a half. Moreover, if the neutrinos have a small amount of magnetism, the solar magnetic field might convert them into undetectable antineutrinos, which would explain the relation to the solar cycle.

When these experiments, and others soon to be developed, have been run for many years, we should be able to sort out all the effects and test the theories of the solar interior. Perhaps more importantly, astronomy and the Sun have given new insights into the nature of the atom and the subatomic particles associated with it, in a wonderful symbiosis between sciences.

ON THE MAIN SEQUENCE

Now we look out into space to see myriads of other stars. Since the Sun is an archetypal dwarf, the main sequence ought to be a hydrogen-burning sequence, and if the theories that are applied to the Sun are correct, they should fit the other stars as well. Over 50 years ago Henry Norris Russell and H. Vogt demonstrated that mass and chemical composition were all that mattered in determining stellar structure and hence luminosity. The chemical compositions of stars are similar. The Russell–Vogt theorem is, then, the theoretical counterpart of the empirical mass–luminosity relation discussed in Chapter 3. As we proceed down the main sequence from the Sun, the mass and the gravitational pressures decline, and so do the internal temperature and the luminosity. The p–p reaction proceeds ever more slowly. At 0.08 $M_\odot$, core temperatures are down to 5 million degrees; any cooler and fusion can no longer be sustained. We have reached the bottom, the lower limit to stellar masses. Our own planet Jupiter, although made largely of hydrogen, is too small to be a star by a factor of 50. However, when we go upward in mass from the Sun and the interior temperature and luminosity increase, something quite different begins to take place. Just after the p–p reaction had been announced, Bethe realized that if the temperature is sufficiently high, another reaction is possible, one that uses carbon as a nuclear catalyst. In normal galactic stars, about one atom of every 2000 is carbon. At lower temperatures, below about 15 million degrees, the atomic velocities are too low for carbon to be penetrated easily by a proton. However, the probability rises rapidly as temperature climbs, so that above this limit a proton is *more* likely to combine with carbon than it is with another proton, initiating the carbon cycle.

When a ^{12}C nucleus merges with a proton (after a typical wait of 13 million years), it has both its atomic weight and number increased by one and becomes ^{13}N. This isotope of nitrogen is radioactively unstable. After about seven minutes one of the protons suffers an inverse beta decay, converting itself into a neutron with the ejection of a positron and a neutrino. The mass stays the same, but the atomic number goes down by one, and the ^{13}N becomes ^{13}C, which is as stable as ^{12}C. The positron and an electron meet, and a gamma ray is again created. After a wait of a few million years, the ^{13}C will grab another proton with the production of another gamma ray, making stable ^{14}N, the stuff we breathe. Yet another proton smashes into the ^{14}N, producing radioactive ^{15}O, which after a minute or so decays into stable ^{15}N. So far, three protons have been consumed. A fourth makes its entry by tunneling into the ^{15}N. Now, instead of the creation of stable ^{16}O, nature offers a surprise. The reaction causes the ejection of a helium nucleus (in the old jargon, an alpha particle), which re-creates the original ^{12}C. So, four atoms of H produce one of He, and the carbon remains unchanged. The Sun is hot enough to produce about 10 percent of its energy by the carbon cycle, and the process must be taken into account in the solar model. The carbon and p–p cycles generate roughly the same amount of energy at about spectral class F0 on the main sequence. Above that, the carbon reactions dominate, and the p–p chain is an inconsequential contributor among the B and O stars.

The basic model of the Sun and stars now also tells us the *upper* limiting mass on the main sequence. As mass and luminosity rise, so does radiation pressure. At 120 or so solar masses, the outward shove of a star becomes so great that gravity can no longer hold it together. Any body that were to form larger than that limit would immediately be disrupted and literally cut down to size. Even at lower masses, the luminous dwarfs have enormously strong winds, as their tremendous radiation shovels quantities of mass into space in stellar winds up to nine orders of magnitude greater than the Sun's.

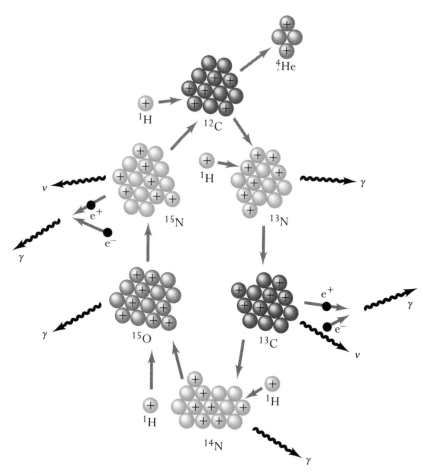

In the carbon cycle, sometimes called the carbon–nitrogen or CN cycle, carbon acts as a sort of catalyst that allows four hydrogen atoms to combine into helium. At all but the last step a gamma ray is produced either directly or by the creation of a short-lived positron.

Theoretical models show that stars of different masses undergo great changes as we proceed up and down the main sequence. Stars more luminous than the Sun have thinner convection zones in their envelopes, and from class A upward, envelope convection disappears altogether, followed by the advent of core convection. Downward through classes G and into K and M the convection thickens. At the bottom of the main sequence it is possible—though not proved—that the convection takes over the star completely, circulating gases from all over into the hydrogen-burning core.

Since convection helps generate the magnetic fields of stars, we would expect these changes to have great effects on what we observe, and we are not disappointed.

BROWN DWARFS

The mass range of stars goes from 120 $M_\odot$ to 0.08 $M_\odot$. The upper limit appears absolute; the lower limit does not. Below this value a star cannot sustain itself by thermonuclear reactions, but that does not mean that such submassive objects—called brown dwarfs—cannot form. Their existence is potentially important to theories of star formation and to the assessment of the Galaxy's mass.

Even though they do not stably burn hydrogen, the brown dwarfs should be visible, radiating dimly first as they burn their small original supply of deuterium to helium and then as they convert their gravitational energy into heat. Since the number of stars increases precipitately down the main sequence, we might think that there should be huge numbers of brown dwarfs, but the search has been daunting. After years of looking, only a small number of candidates have been uncovered. One was found by its very low surface termperature of about 1000 K. A handful are suspected from the existence of atmospheric lithium, an element easily destroyed at normal stellar temperatures. And about a dozen are inferred from the gravitational effects they have on stellar companions, from which we estimate mass, the candidates themselves unseen. In spite of their bulk popularity, the number of M dwarfs seems to drop off cooler than M5 or so, the rare brown dwarfs apparently continuing the trend.

STELLAR ACTIVITY

There is no reason to believe that the Sun is unique. Other stars should have the same sorts of bizarre characteristics: the granulation, spots, magnetic fields, activity, flares, and the like. There are probably few astronomers alive who have not dreamed of actually seeing the disks and surface features of other stars, first just for the wonder of it, and second, because the exterior properties provide powerful constraints on the internal structures. But the inscrutable stars give us little satisfaction. Even the angularly largest of them is only a few hundredths of a second of arc across, and since our turbulent atmosphere gives at best a resolution of about a half a second, the disks—with the one exception illustrated in Chapter 3—remain hidden from view.

Nevertheless, astronomers have learned a great deal by taking more indirect routes to knowledge. We have been able to detect and study stellar chromospheres and coronae, have found starspots, and have even seen active stellar phenomena that mimic—and sometimes violently exceed—those of the Sun. The solar chromosphere and corona are overwhelmed by the brilliant photosphere, and in integrated light (radiation summed over the whole surface) they would be very hard to see. But careful spectrographic observation, with which we can increase the contrast, allows the detection of stellar chromospheres, which appear from class F down to and including M. In these stars, the photospheric H and K absorptions produced by ionized calcium are enormously powerful. At the centers of these lines nearly all the photospheric radiation is blocked, and the stars become extremely dim. However, the chromospheres, consisting of gas under low pressure, produce *emission* lines at H and K, which can then easily be seen against the darkened background. As we proceed down the main sequence from the Sun, the ionized calcium emissions strengthen relative to the integrated stellar luminosity, showing that the chromospheres become thicker, apparently in response to the growing magnetic fields generated by the deeper and deeper convection zones.

We know that the magnetic dynamo is produced by a combination of convection and *rotation*. Remarkably, we can determine the spin rates of these tiny specks of light with considerable accuracy. An absorption line is generated from all points on the surface of a star. If a star is in rotation (and if we are not looking right down on the pole), some of it must be coming at us (relative to the overall radial velocity) and some must be going away. Therefore part of the absorption is Doppler-shifted to shorter wavelengths relative to the average and part to longer, resulting in the broadening of the line. The profile of the absorption—its functional form relative to wavelength—takes on a very characteristic shape, from which we can deduce rotation velocity. Since the rotational axis of a star is rarely placed exactly perpendicular to the line of sight, what we actually derive is the projection of the velocity against the plane of the sky. By assuming that stellar axes are randomly oriented, we can find statistical averages for specific spectral classes. Activity in the stellar chromospheres also provides information. We can actually see the H and K line emissions change periodically as active regions move in and out of sight, allowing the rotation period to be determined directly.

We find that there is a strong correlation between main sequence spectral class and rotation speed. The G, K, and M stars all turn slowly, the

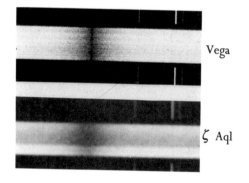

The hydrogen lines of Vega (top) and Zeta Aquilae (bottom): Zeta Aql's broadened line shows it to be rotating with a speed of 345 km/s as compared with Vega's mere 15 km/s. We are only detecting rotation against the line of sight; Vega appears actually to be a rapid rotator when viewed pole-on.

Vega

ζ Aql

speed dropping from about 5 km/s to 1 km/s or so near the end of the main sequence. But as we enter class F from below, we encounter what is called the rotation break, and suddenly the spin rates jump to 100 km/s and up to over 200 km/s in class B. The break occurs on the main sequence about where envelope convection ceases. Furthermore, rotation is closely associated with stellar age. As we will see in the next chapter, the theory of stellar evolution allows us to tell the ages of star clusters. Stars in older open clusters are always spinning more slowly. The rotation combined with the convection zone produces the magnetic fields and stellar activity, *and then the magnetic field applies a brake that actually slows the star.* As mass is lost from a star, as it is in the solar wind, it carries the magnetic field lines with it. We know this to be true from observation of the Sun. But the field lines are still anchored to the star, and slowly, over billions of years, they act like ropes that drag on the rotating ball of gas, gradually bringing it to a near halt. Nevertheless, activity stays strong in the cooler stars, showing that as long as there is a deep convection zone, not much rotation is needed to produce a significant magnetic field.

It is unfortunate that we cannot all spend a few hours at the telescope watching a few faint red dwarfs. These dim stars are generally so unassuming that we pay them little attention. A substantial number, however, exhibit flares that put those of the Sun to shame. It is rare to see a solar flare that is visible through the telescope in white light. We normally have to look in chromospheric emissions like the H and K lines or Hα to see them. Not only are flares visible from M dwarfs, they can increase the brightness of the star by one or several magnitudes. Imagine the Sun suddenly becoming two or four times brighter for a few minutes! Like the solar variety, these flares are also visible in the radio, ultraviolet, and X-ray parts of the spectrum. They are so powerful that they may involve magnetic release over the whole star. Flare activity is clearly related to rotation and age, as young clusters contain many more active flare stars than do older ones. It is a fascinating exercise to consider what our own Sun may have been like billions of years ago when life was first developing and when it was young and spinning more rapidly. What effect could enhanced activity have had on the evolution of living things?

Where there are flares there ought also to be coronae. We would not necessarily expect a stellar corona to be like that of the Sun, but it should certainly be a fascinating variation on the theme. We know that coronae exist around other stars from simple quiescent observation of their X-rays by Earth-orbiting satellites.

Activity seems to go berserk in certain kinds of binary stars. The RS Canum Venaticorum stars are close pairs of G dwarfs or subgiants, tidally

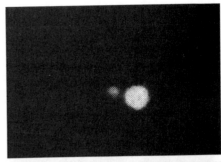

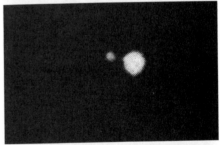

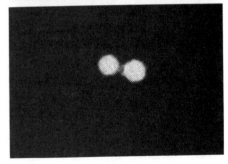

Photographs of DO Cephei, a binary consisting of M3 and M4 dwarf components, dramatically show how luminosity drops along the main sequence. Suddenly, and without any warning, the dimmer M4 dwarf pops a powerful flare and brightens by a full magnitude, a factor of 2.5. The degree of brightening is much greater in the ultraviolet.

locked onto one another in much the way that the Moon keeps one face pointed toward the Earth. Quick orbital revolution has produced rapid axial spins, which together with the stars' deep convection zones produce awesome activity. They are so covered with starspots (it is the only explanation) that we can actually see the whole star vary significantly as active regions turn in and out of sight. They also display prominent coronae through X-ray radiation and occasionally pop enormous flares, solar-type activity gone mad, clearly demonstrating the role of rotation in generating the magnetic dynamo.

On a quieter level, if we look closely enough at solar-type stars, we see evidence for entire stellar magnetic cycles that are quite reminiscent of our own of 22 years. About two-thirds of the stars examined display regular variations with multiyear periods. It is tempting to relate the third that show nothing at all with the times when the solar cycle shuts down, as it did during the Maunder minimum.

Such long-term studies are some of the most difficult in astronomy. Telescope time is precious, and the pressure to publish results is great. Anyone examining the problem of stellar cycles must have enormous patience to follow a star over intervals of decades—yet the payoff is huge. Eventually, we will be able to correlate stellar activity with other observational parameters, giving us a much better understanding of the phenomenon. The connection between other stars and the Sun may then allow us to know our own star better, the one that we depend on for our very lives.

UP THE MAIN SEQUENCE

Above class F the convection disappears, and we might think that stellar surfaces would be more placid, if not even boring. Again, we are in for a delightful surprise. Zeeman splitting of spectral lines shows that A dwarfs can exhibit magnetic fields far more powerful than that found in the Sun or even in the biggest sunspots. The record, held by dim ninth-magnitude HD 215441, is a general field some 110,000 times the strength of Earth's! Regular variation in the field strengths of these Ap (A-peculiar) stars clearly shows that the magnetic axis is tilted relative to the rotational axis, an effect commonly seen in the magnetic fields of the planets. There is a dynamo of some sort at work, but as yet we do not know what it is.

These stars also display weird abundance anomalies, showing overly strong lines of silicon, chromium, strontium, and even the rare earths such as europium. The line strengths are variable along with the magnetism.

|← 19 Å →|

Cr II 4558

The spectrum of HD 215441, a ninth-magnitude star in Lacerta, displays ionized chromium lines in the violet that are split by the Zeeman effect, indicating a magnetic field 110,000 times that of Earth.

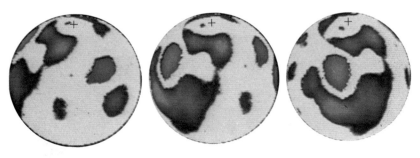

By carefully measuring the way in which the shapes of silicon lines change in the Ap star γ^2 Arietis, astronomers were able to reconstruct how spotted areas that contain strong overabundances of silicon move in and out of sight. We therefore see computer-generated views of the stellar surface itself. The green areas show silicon overabundances, the red areas depletions. The "+" marks the rotation pole.

Apparently, there are patches vaguely reminiscent of solar active regions that are highly enriched in particular elements and that rotate in and out of sight. The best explanation is that some kind of diffusion is at work that can cause some elements to settle downward out of the atmosphere and others to rise. The whole matter is a considerable mystery.

Some type of diffusion may also be at work in the metallic line or Am stars. These show no significant magnetic fields, are underabundant in such elements as calcium, and strongly overabundant in heavy metals like strontium, yttrium, and barium. There is also good evidence that they are in binary systems, scoring one more effect for stellar duplicity.

It seems the only time we can say that we "understand" the stars is if we do not look too closely. The wealth of detail they exhibit is overwhelming, and the closer we look the more we see to amaze us.

The Dumbell Nebula.

Chapter Five

COMING OF AGE

THE DEVELOPMENT AND AGING OF THE SUN
AND THE LOWER-MASS STARS

The old astronomers spoke of the "fixed stars," which, unlike the wandering planets, apparently never change their course across
the celestial sphere. Watch all your life, you will see no deviation. Our
constellation patterns, seemingly unaltered over thousands of years, pro-
claim the constancy of the heavens. In Chapter 3, however, we saw that
the stars *do* move and that, given long enough, the mythological beasts
and beings that populate the sky will all disappear. Now we address a
different kind of change. Stars live long lives, the Sun 10 billion years,
but none lasts forever. Their energy supply is limited, and when it runs
out they—like the forms of the friendly constellations—will also die.

Something so large and powerful as a star is not going to go quietly. Stars do not just suddenly wink out but go through death throes that can produce massive swelling, wild variations, devastating explosions, literal punctures in space, and some of the most beautiful sights of all nature. This activity, neither subtle nor hidden from view, is in the sky for all to see, for it is the brilliant giants and supergiants that are departing.

The death act is quick in a stellar context but, with some singular exceptions, it is creepingly slow in human terms. Astronomers have to look at the entire array of celestial sights and string them together with the aid of theory to see how the quiet dwarfs of the main sequence transform themselves into the wonders of the celestial sphere.

STELLAR LIFETIMES AND THE REALMS OF THE MAIN SEQUENCE

The main sequence is the broad band of stellar youth and middle age where stars lounge while burning their initial supplies of hydrogen into helium. Earlier, we saw that stellar temperatures and luminosities are controlled by mass. We will now see that mass commands even how long the stars live and how they will mature and finally expire.

Given two piles of wood and a match, you would naturally expect the bigger pile to burn the longer. Apply this reasoning naïvely to stars and you will be utterly wrong. More massive stars clearly have larger energy supplies; whether the fuel is gravity or nuclear power is at the moment irrelevant. However, in 1926, Eddington showed that in order for a star to maintain equilibrium, the luminosity must be roughly proportional to M^3, in rough agreement with the mass–luminosity relation derived from binary stars. This relationship allows an examination of stellar lifetimes. Look at a simple argument that ignores all the subtleties. Crudely, the lifetime of a star, τ, must be proportional ($\propto$) to its energy supply divided by the rate at which it is used. Since energy supply $\propto$ mass and the rate $\propto$ luminosity, $\tau \propto M/L$. But since $L \propto M^{3.5}$ (the average relation from binaries), $\tau \propto M/M^{3.5}$, or $\tau \propto 1/M^{2.5}$. On the basis of this simple relation, Vega, a typical A dwarf 2.5 times the solar mass, will live $(1/2.5)^{2.5} = 1/10$ as long as the Sun, and a B dwarf of 10 solar masses will expire in only 1/300 of that time. At a tenth of a solar mass, down near the bottom of the main sequence among the dim red M dwarfs, stars use their meager fuel supply so parsimoniously that they will endure 300 times longer than the Sun. More precise values can be calculated by factoring in the actual value of the exponent in the mass–luminosity relation, which changes some with the mass.

Absolute stellar ages, as we saw with the Sun, depend on the nature of the fuel. Nuclear fusion allows the Sun to live on the main sequence for 10 billion years. If we combine this number with the above proportionalities (using the variable exponent in the mass–luminosity relation), we find that a rare O dwarf at the top of the main sequence will survive for a mere 3 or 4 million years—here today, gone tomorrow. *Some stars have come and gone since our human race has walked the Earth.* Down in the depths of the main sequence, a cool M dwarf will last for an astounding *3 trillion years.*

The main sequence ends at 0.08 M$_\odot$ (class M8), below which we find the fleeting brown dwarfs of Chapter 4, ''nonstars'' that are only now being discovered. Below 0.8 M$_\odot$ (class G8), lifetimes are longer than the age of the Galaxy, which we believe to be about 13 billion years (how this number is known is a principal subject of this chapter). None of the stars between these limits, those within the lower main sequence, has yet had time to perish—every one that ever was *is still there.* They consequently play no further role in any meaningful discussion of evolution. Remarkably, because the number of stars rapidly increases down the main sequence, this realm includes some 90 percent of all dwarfs.

The intermediate main sequence, though far less populated, is much more interesting. Its range is between 0.8 and *about* 8 M$_\odot$ (spectral class B3), and thus includes our Sun. The upper limit is poorly known: It might be 6 M$_\odot$ (B5) or it might be 10 M$_\odot$ (B2). This domain includes some 10 percent of the dwarfs and is the origin of all the giants like Arcturus, Aldebaran, and Mira, and of the white dwarfs such as 40 Eridani B and Sirius B. These main sequence stars, though they undergo some spectacular changes, die with quiet dignity.

Upper main sequence stars, those between about 8 M$_\odot$ and the maximum allowable value of about 120 M$_\odot$, are rare, constituting less than 1 percent of the total population. These are ephemeral beings, lasting at most a few tens of millions of years. Since they are the sources of the supergiants, their enormous luminosities and evolutionary effects make them famous out of all proportion to their numbers. Apart from the difference in lifetimes, the great distinction between these stars and those of the intermediate main sequence is that they are too heavy to produce white dwarfs. Some may expire quietly, but the majority (perhaps all) do not. Instead, they are wiped out in sudden blasts called supernovae, in the process creating some of nature's most bizarre objects. In Chapter 6 we will wander through the wonders of the upper domain. Though scarce, the

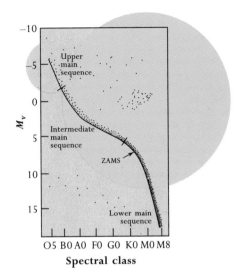

Stars on the lower main sequence—61 Cyg, Barnard's Star, Proxima Centauri—have never had time to evolve. An intermediate-mass star like the Sun or the A dwarfs Sirius and Vega will, in a few billion years, evolve first into a giant, then a white dwarf. The young massive stars of the upper main sequence, which mark the Galaxy's spiral arms, are supernova candidates. The left-hand edge of the distribution is the zero-age main sequence, or ZAMS, where stars begin fusing a fresh supply of hydrogen. The density of points is schematically proportional to the number of stars of each kind per unit volume. The giant branch begins at class G on the main sequence and goes up to the right. Typical main sequence and giant diameters are superimposed.

great beasts who live there have a profound impact on the Galaxy and on the intermediate mass stars. Here we will look at the intermediate zone—at stars like the Sun—to learn our own fate.

EVOLUTION ON THE MAIN SEQUENCE

All of stellar evolution can be understood from one simple rule: A star is always under the influence of its own gravity and is trying to make itself as small as possible. For many reasons it will be prevented from doing so for various periods of time. All the stages along the way can be identified as either pauses in the shrinkage or as the acts of the shrinkage itself.

A star begins by collapsing out of the turbulent mixture of interstellar gas and dust. The first stop in its life history is by far the longest, the main sequence. Brand-new stars lie along the left-hand edge of the band, on a locus called the zero-age main sequence, or ZAMS. Here the temperatures in the cores of the contracting bodies become high enough to sustain thermonuclear fusion. The burning of hydrogen into helium creates so much internal pressure that the great squeeze is temporarily halted, and the star stabilizes.

Much like human beings, however, stars are always aging and evolving, even on the main sequence. It is a matter of degree, of the rate at which transformations occur. In early adulthood we change relatively slowly, but old age moves in quickly, producing ever-accelerating alterations. From the time it is born, a star begins using its finite fuel supply, and therefore must change. The close balance in the core between the diminishing fuel supply and the increasing burning rate maintains stellar luminosity. Without this stability stars would change quickly, and there would be no main sequence. Good as it is, however, the balance is not perfect. The ever-increasing temperature actually wins out a bit over the diminishing fuel, and the outer part of the core expands slightly into the envelope, where fresh hydrogen awaits. The result is that the luminosity of the star slowly *increases* with time. The surface of the Sun now glows with an effective temperature of 5780 K, but 5 billion years ago, when it was born and had a full complement of hydrogen (90 percent H, 10 percent He), it was somewhat redder and cooler, closer to 5500 K. It was also smaller and produced only about 70 percent of its present radiation. We have now used about half of the core hydrogen. Over the next 5 billion years, the rate of change will accel-

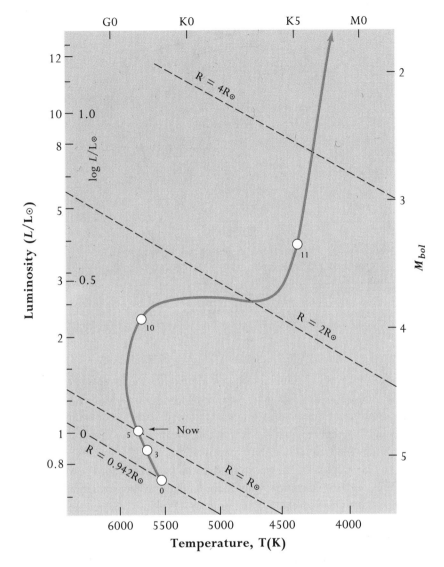

Theoretical calculations of the early evolution of the Sun show us that at birth, with its full hydrogen supply, it was 94 percent its current size, about 70 percent as bright, and a bit cooler at the surface. It will continue to heat and brighten slowly over the next 5 billion years. When the hydrogen at the center runs out, it will move into the subgiant realm of the HR diagram on its way to becoming a giant. The form of this graph differs from the HR diagrams presented earlier: Here luminosity is plotted directly against temperature (actually, their logarithms), instead of magnitude against spectral class.

erate. When the hydrogen finally runs out, our Sun will be twice as bright as it is now and 75 percent larger. Slow changes such as these help give the main sequence its breadth, making it a band on the HR diagram and not a line. It is further broadened by differences in composition that shift the subdwarfs of Population II to the left.

What do these changes mean to life on Earth? A drop of 30 percent in the luminosity of the Sun would be devastating to us—water would freeze

Stellar masses are indicated on the zero-age main sequence. Below 2 $M_\odot$ an evolving star moves slightly to the right and then brightens into a giant. The ascent is terminated by the helium flash. As it begins to burn to carbon, the star descends the giant branch to the clump. If the metal content is low, these helium-burning stars will spread out to the left along the horizontal branch according to their masses. If they land in the instability strip they become RR Lyrae variables. As a 5-$M_\odot$ star grows to gianthood, it cools to the right at a fairly constant luminosity; when it begins to burn helium at the red giant tip, it loops back. During evolution, it becomes a Cepheid twice. Theoretical HR diagrams like this plot the logarithm of luminosity against that of temperature, distorting the linearity of the spectral sequence. Bolometric magnitude is at right.

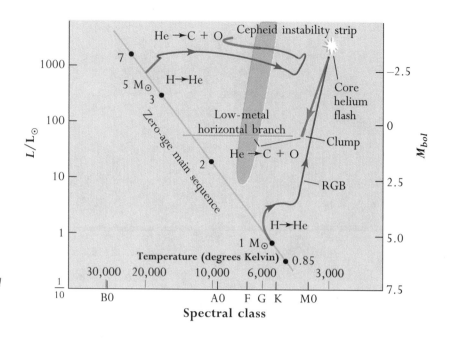

and the planet would become more like Mars—yet the fossil record shows that the conditions for life here have remained hospitable over the ages. Perhaps our atmospheric blanket was different earlier, containing more carbon dioxide that trapped the lower solar heat. The Sun must also have been spinning considerably faster when it was younger, before its magnetic brake had been applied. The magnetic dynamo had to be more powerful, so that flares and the accompanying X-ray and ultraviolet radiation should have been at higher levels. We may never fully know the reason.

Our Sun's future, though, is predictable and inevitable. It will become much brighter. Within 2 billion years, or perhaps far fewer, there will be little in the way of winter. Higher temperatures will first cause greater evaporation of the oceans and increased atmospheric opacity. The temperature will climb still further, leading to a runaway greenhouse effect and conditions more like those found on Venus. Solar ultraviolet radiation will then break down the water molecules, and their hydrogen will escape forever. Even if our atmosphere were to stay stable, which it almost certainly cannot, by the time the main sequence phase is over the Earth will be so hot that life will be long gone. But such events are far in the future. Considering that humanity has been on the planet for a period measured in only millions of years, and that we have written records for only thousands, the thought should not be too depressing. We have a long way to go.

GIANTS IN THE SKY

Main sequence evolution is but a prelude, its stars remaining relatively stable until nearly the entire central fuel supply is gone. Then the *real* action begins. Look first at the more populous stars, those like the Sun. When nuclear reactions finally cease deep in the core, there is no longer any support against the inexorable force of gravity. The star has waited 10 billion years to use its vast supply of gravitational energy, and now the time has come: The core must rapidly (at least in astronomical terms) contract. The outer part of the core, where it has been cooler, still has some hydrogen left, and nuclear reactions continue in a shell, which now grows into the envelope. The changes in internal chemical composition, energy-generation rate, and opacity begin to produce massive alterations in the star's internal and external structure. For a period equal to about 20 percent of the main sequence lifetime, luminosity stays fairly constant as the radius nearly doubles and the surface cools to under 4500 K. Then evolution speeds up. The point that represents the star on the plane of the HR diagram begins to move rapidly across its surface to climb the evolutionary giant branch, the path that takes it into the realm of the giant stars. In only a bit over half a billion years the brightness shoots to over 1000 times the current solar value. The outer envelope expands enormously, causing the effective temperature to drop to 3500 K, and there is a new M0 giant in the sky, blazing forth at an absolute bolometric magnitude as bright as -3. Somewhat more massive stars follow the same general course, becoming even brighter—another magnitude or so—in shorter times. A 2-$M_\odot$ star will make the transition after the cessation of core H burning in only 300 million years, and one of 3 $M_\odot$ in but 100 million.

The effect as viewed from an orbiting planet—could anyone survive to see it—would be awesome. Our Sun will increase its size perhaps 100-fold to an AU in diameter (it is now 0.01 AU), pushing past the orbit of the planet Mercury. Reddish and bloated, it will appear 50° across in the Earth's sky, quadruple the angular size of Orion, and (if we ignore the inevitable slowing of the terrestrial rotation) will take over 3 hours to rise and set. The new giant will now be seen with the naked eye to a distance of over 500 parsecs (1900 light-years).

As it matures, the surface gravity goes down, and the once-quiet wind from its surface increases its strength a million times or more. The star is beginning one of its greatest acts: It starts to whittle itself away, returning its mass to interstellar space, its birthplace.

As the Sun expands something strange begins to happen that presages the end product of evolution. Remember the uncertainty principle—that

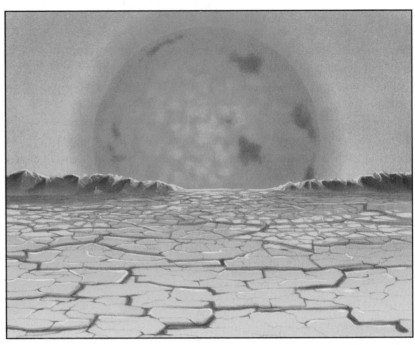

A swollen Sun—half the size of the orbit of Mercury—rises as a red giant to blister Earth.

the uncertainty in momentum times uncertainty in position approximately equals h? The Planck constant, h, represents something of a minimum in dimensionality. If you cube it you have the unit volume of a six-sided space called phase space that has three dimensions in momentum and three in real space. In 1925, Wolfgang Pauli formulated the exclusion principle, which states that within a unit phase space no two identical electrons can exist. Like all subatomic particles, electrons have a property called spin whose value must be plus or minus half a unit, the different signs indicating spins of opposite directions. Therefore, within the volume h^3 there can be at most two electrons with opposite spins. You cannot squeeze the electrons together any tighter. (One result of the exclusion principle is that no two bound electrons can occupy exactly the same orbit, so multiple electrons are forced into shell structures.)

The disparity in size between the envelope and the helium core of a fully developed giant reaches astounding proportions. As the outer parts expand to interplanetary dimensions, the core contracts toward the size of Earth. The density gets so high, in the neighborhood of a million grams per cubic centimeter, that the lower-velocity electrons fill their phase spaces. The gas is now said to be degenerate. If any electrons are added, they can exist only at high velocity, where phase space still has room. The perfect gas law breaks down, and pressure depends on density alone and no longer on temperature—that is, temperature can increase while the pressure is constant. The core now has the characteristics of a white dwarf.

When the interior temperature reaches 100 million degrees, what was the ash of the old reaction, the helium, suddenly and explosively begins to fuse and becomes a new fuel in an event called the helium flash. Helium does not burn easily. If two helium atoms collide, they produce ^{8}Be (the normal isotope of beryllium is ^{7}Be), which is highly unstable and on average falls apart back into the helium pair in only 10^{-16} seconds. However, there must be an equilibrium in this reaction, so that at any one time there will always be a tiny amount of ^{8}Be. If one of these ephemeral atoms can collide with another helium nucleus, the two can fuse to create stable carbon, ^{12}C, and a gamma ray. Such a reaction requires that three helium nuclei (α

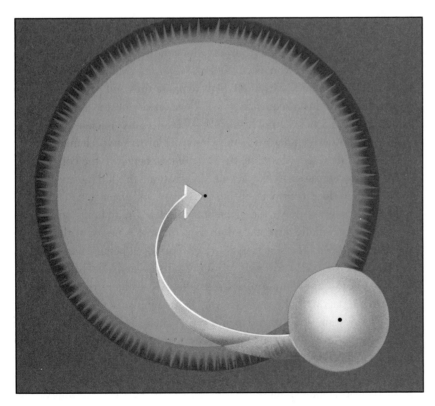

The degenerate core of a fully developed giant star—invisible on this scale without magnification—can just be seen on the right, where the dot in the center is blown up.

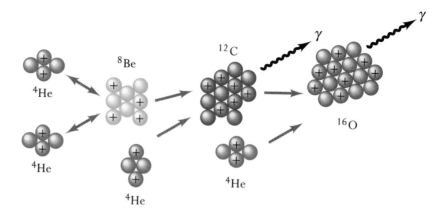

The 3-α reaction, described by Edwin Salpeter in the early 1950s, involves the near-simultaneous collision of three helium nuclei (α particles) to produce carbon, ^{12}C, and a gamma ray. The ^{8}Be nucleus formed by the collision of a pair of α particles decays almost instantly; the third nucleus must come along at just the right time. An occasional collision with yet another α particle will produce ^{16}O and more gamma radiation.

particles) come together practically simultaneously, and the event is known as the 3-α process. It can occur only if vigorous collisions happen very frequently and requires extremely high temperatures and pressures, which are unavailable in main sequence stars. The necessary heat is in fact so great that another α particle can occasionally penetrate a ^{12}C nucleus, producing ^{16}O (plus another gamma ray) and a mixture of the two elements.

With a new energy source, the state of degeneracy in the core is lifted and the core expands. In the act of stabilization, the luminosity actually drops, the envelope contracts, the surface heats, and the star descends to a point about halfway down the giant branch. Here, in a clump on the HR diagram, it will quietly burn its helium in a core that is nested within a shell of fusing hydrogen. The star is in a sort of short-lived helium-burning main sequence. Many well-known K giants, Arcturus and Aldebaran among them, are in this configuration, as are the four K giants of the Hyades HR diagram and many of the stars seen in the magnificent globular clusters.

The location of a dwarf depends on its chemical composition, and so does that of a helium-burning star. Those that are deficient in metals, the ones of the halo and of the globular clusters, will be displaced to the left of the clump. Small differences in mass will then cause them to spread into the distinctive horizontal branch. The ones that fall into the instability strip become the RR Lyrae variables, the low-amplitude oscillators with short (under-one-day) periods we met in Chapter 3.

The stars along the upper part of the intermediate main sequence, above 4 $M_\odot$, go through some of the same internal changes, but their behavior is somewhat different. Instead of a short cooling track on the HR diagram followed by a great rise in luminosity, they undergo a long period of cooling from class B or so to class K and then a smaller rise in luminosity. They populate much of the realm of the bright giants in the HR diagram, luminosity class II, and swell considerably, increasing their already large sizes by a factor of 50 or so. Following helium ignition, which is no longer explosive, they are not very stable and move back and forth on the diagram as they alternately heat and cool. The transition from the lower part of the intermediate main sequence to the upper is not sudden but gradual, a continuum of behavior.

As the higher-mass stars move across the HR diagram, they also encounter the instability strip and are seen as Cepheids. This narrow zone is produced when layers of hydrogen and helium ionization deep within the stellar envelope trap and valve the outgoing radiation. When they hold it, the star shrinks. The shrinkage in turn changes the internal temperature structure, which alters the ionization balance, releasing the radiation—and the star expands. The result is a pulsation that makes the star alternately brighten and dim. The more massive stars are more luminous and larger and take longer to throb, explaining the period–luminosity diagram of Chapter 3 and the short periods of the less massive RR Lyrae stars.

The evolution to this point is impressive, to be sure, but it is only an overture. Grander things will come when even the helium runs out, and the stars begin to pace their last miles to expiration.

EVOLUTIONARY AGES AND THE GALAXY

Now the HR diagrams of star clusters like the Hyades, Pleiades, and the globular M 5 start to make some sense. A cluster is born with a full range of masses that occupy the entire main sequence. The Double Cluster in Perseus (h and χ Persei) is an example, a uniquely linked pair of systems born just yesterday whose main sequence runs all the way to the O stars. As a cluster ages, the massive stars die first, and the main sequence begins to disappear from the top down: The stars on the upper main sequence become supergiants, those on the intermediate section, giants.

With the aid of theory we can actually date a cluster by observation of where the dwarfs stop: h and χ are about 2 million years old. The Pleiades cluster is older: It has no O stars, but it does have B stars, showing it to

NGC 3293, an open cluster of stars, has most of its main sequence intact.

Alan Sandage's comparison of galactic, or open, clusters (where the stars are plotted against color instead of spectral class) shows vastly different main sequence termination points. Clusters with intact upper main sequences are very young. Stars peel off the main sequence from the top (the most massive) downward as they evolve into supergiants and giants. The globular cluster M 3 looks as if it is younger than the oldest open cluster, but that is an effect of low metallicity, which moves stars to the left on the HR diagram.

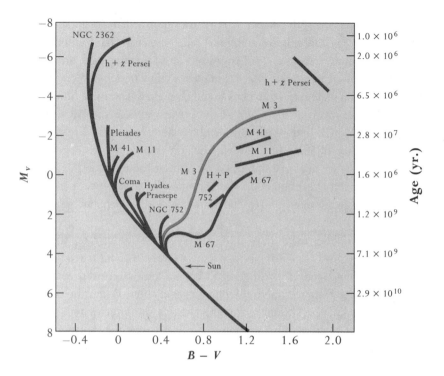

have an age of about 10^8 years. Its half-sisters, the Hyades, with no main sequence stars brighter than class A, are older still—born some 10^9 years ago. M 67 in Cancer, at around 5 billion years, is one of the older open clusters known, with a main sequence that stops just brighter than the Sun. As a cluster ages, its giants keep pace with the main sequence and also descend in luminosity. Stellar evolution is fast enough so that the observed locus of giants on the HR diagram approximately outlines the path of evolution.

Applying the theory of evolution to clusters provides a powerful way to study the Galaxy. We can date the age of the Earth and solar system from the oldest rocks, and we can tell the age of the Galaxy from the oldest clusters. The open clusters occupy the galactic disk exclusively, so we know that the plane of the Milky Way must be at least as old as the oldest of them, about 8 billion years. Unfortunately, open clusters are loose affairs. Even the richest contains only a few hundred or a thousand stars, and the internal gravitational forces that bind them together are not very strong. Over time, internal interactions fling stars away, and tides raised by the dense mass of the Galaxy cause further disruption. Most open clusters are rather young; it is rare for one to survive even as long as M 67 has. Consequently, these assemblies give only a lower limit to the disk's age.

Now we can understand the general HR diagram of the stars of the Galaxy as well. The giant branch is densely populated with stars, but bright giants and supergiants are scarce. The vast majority of stars that we see locally, those that appear on the HR diagram, are in the disk. They are being born all the time and have a variety of ages and masses. All the stars above G8 or so on the main sequence are able to produce giants or supergiants. The luminosity function, however, increases dramatically down the main sequence, so the space density of supergiants and giants must also strongly increase downward toward lower luminosity. The evolution of the lower-mass stars provides the large bulk of the giants, and we see a distinct, heavily populated branch streaming up and to the right from the middle of the main sequence. In principle, we could derive the age of the disk as we do the clusters, but the observations are too confused by a spread of masses, ages, chemical compositions, and stars mixing in from the thick outer disk and even from the inner halo. Nevertheless, from noting where the general giant branch intersects the main sequence, we can tell that the disk is at least as old as the oldest open clusters.

We fare better with the globular clusters. The majority are in the Population II bulge and halo. They are packed with stars and are so tightly bound together that they defy easy disruption. The best fit between theory and observation shows that the oldest of these magnificent groups were born about 14 to 16 billion years ago. They are older than the open clusters and clearly older than the stellar assembly within the disk. Since we find nothing of greater age, we identify that number with the age of the Galaxy itself. Because they are part of the constituency of the galactic halo, it is evident that the halo was the original component of the Galaxy and that the disk, with its current complement of O and B stars, came later. The set of noncluster halo subdwarfs, whose main sequence above class G long ago burned away, yields the same picture.

Since the globulars and subdwarfs have fewer metals than the disk, the number of heavier atoms clearly must somehow be increasing as the Galaxy ages. Stars are producing them—but how? We have already seen how helium and carbon are manufactured as by-products of stellar power generation. The general problem was solved in one of the landmark papers of twentieth-century astrophysics, published in 1957 by E. Margaret Burbidge, Geoffrey Burbidge, William Fowler, and Fred Hoyle (affectionately referred to as B^2FH), who described the various processes of nucleosynthesis, the creation of heavier atoms from lighter ones. As we turn to more

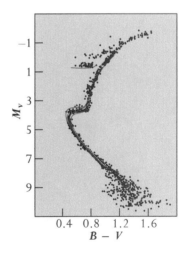

The HR diagram of the great southern globular cluster 47 Tucanae is fitted with a series of isochrones, lines that show the evolutionary distributions at different ages (10, 12, 14, and 16 billion years from top to bottom). The best fit appears to be between 12 and 14 billion years, which allows us to date the formation of the cluster and the halo of the Galaxy.

advanced stages of stellar evolution and, in the next chapter, to the evolution of high-mass stars, we will begin to see how these processes work and how the new atoms arrive at the stellar surfaces.

GIANTHOOD REPRISED

Helium burning sustains only a short hiatus in the inevitable contraction. Once this fire gets started, the reaction rate is so fierce that the fuel supply does not last very long. Soon the last of the helium is gone from the central core, leaving a huge ball of carbon and oxygen, and the shrinkage starts all over again, with the same results. The internal temperature rises, and now the helium burning sweeps outward in a shell around the collapsing defunct core, fed from above by a shell that is still fusing hydrogen. Once again the outer layers begin to swell and cool. The star begins its second ascent of the

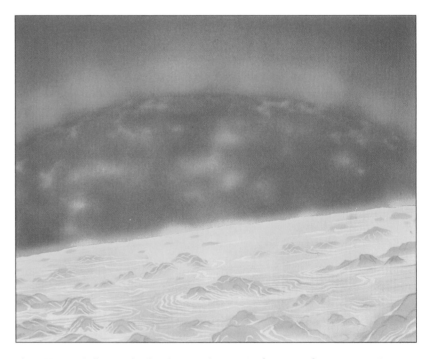

The AGB Sun, brilliant and red, a huge wind screaming from its surface, rises over the Martian landscape—if anything remains of the little planet.

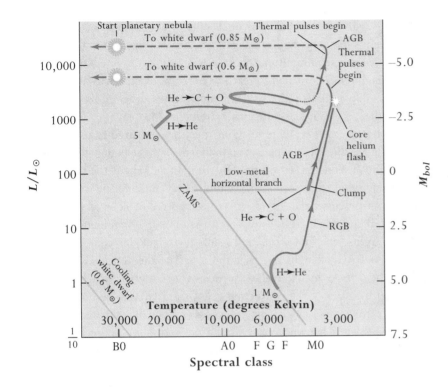

The theoretical HR diagram is expanded here to show the complete evolutionary tracks of stars of 1 $M_\odot$ and 5 $M_\odot$. After they have finished burning helium, both stars ascend the asymptotic giant branch with hydrogen-burning shells. Thermal pulsing, in which helium and hydrogen shells turn on and off in succession, starts near the top. Mira variability also starts at about these points. During the second ascent, the stars furiously lose mass. At the dashed lines most of the mass is gone and the nearly bare cores heat to the left, eventually illuminating the escaping mass to become planetary nebulae. The cores get so hot they move off the diagram, and later appear below as cooling white dwarfs.

giant branch, this time with a more powerful energy supply of not one but two nuclear-burning shells.

The second ascent rather parallels the first. The stars from the horizontal branch climb more or less asymptotically to the tracks taken on the first ascent, and this new phase has come to be known as the asymptotic giant branch, or AGB. (The first ascent is distinguished by calling it the red giant branch, or RGB.) Nuclear burning now proceeds in a strange fashion, the two shells switching on and off in sequence.

As the ascent begins and the star swells, the H-burning zone moves outward, cools, and shuts down, the luminosity coming from a thick shell of fusing helium. As brightening proceeds, the helium begins to burn away. The old H-burning shell contracts and re-ignites, feeding fresh helium into the zone around the still-contracting core. Then, with searing violence, the new helium shell explosively re-ignites, a nuclear bomb buried deep within the expanding star. The H shell retreats and shuts down to wait its turn, which will come when the new helium is gone. These helium flashes or thermal pulses come closer and closer together as the star nears the top of

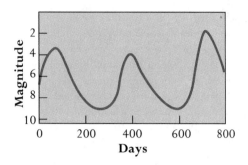

The graph of visual magnitude against time shows fairly regular pulsations with a period of about 330 days as Mira waxes and wanes. The maximum and minimum brightness can change considerably from one cycle to the next.

its great—and last—ascent. So much energy is supplied that the stars become quite luminous and attain immense sizes. The Sun will reach an absolute bolometric magnitude of −4 or so. It is expected to swallow Mercury and Venus both, and to reach to the orbit of the Earth itself. These stars are so swollen that the surface temperatures drop to 3000 or even 2500 K; their spectral classes are as extreme as M8.

To test the theory, its predictions must be compared to observation. By and large, the agreement is good, the theory predicting the correct general distribution of stars on the HR diagram and about the right luminosities. However, there are some "illuminating" differences, which let us know about processes that have not yet been correctly included. The theoretical RGB and AGB tracks show about the same low minimum temperatures; in reality, the first-ascent red giants never get this cool. Moreover, the observed AGB flattens out near the top. The theoretical RGB can be slid to the left to match reality by decreasing internal opacities or by increasing the sizes of the convection cells. The AGB can be brought into better agreement by also changing mass-loss rates. Unfortunately, none of these quantities is known very well, and we can do no more than make uncertain corrections that cannot actually be built directly into the theories.

As an AGB star brightens, its envelope becomes unstable. Somewhat like a Cepheid, it begins to valve the outflowing radiation and starts to pulsate. It is so huge that, unlike a Cepheid's, its oscillation period, instead of being a matter of days or weeks, is months or years. We then look into the sky to see a long-period, or Mira, variable. Mira, *o* Ceti, is the archetypal AGB star. This monster (appropriately found in the constellation Cetus) is the size of the orbit of Mars, and takes 330 days to expand and contract, changing its radius by a factor of 2. The Miras are everywhere and are the most common type of intrinsic variable known. Periods range from a hundred days or so to over five years. The amplitudes, the differences between maxima and minima, can be as much as 10 visual magnitudes. Such large fluctuations are deceptive, however. The change in radius produces a notable variation in temperature. The star cools so much that a good part of its radiation disappears into the invisible infrared, rendering it quite faint. The bolometric amplitude, though still impressive, is much less.

The Sun is expected to swell to the size of the terrestrial orbit. If it overtakes our now-molten planet, the Earth will actually orbit inside the low-density envelope. Friction with the gases will cause it to lose orbital energy and spiral inward, eventually to be utterly destroyed along with Venus and Mercury. Mars will likely be spared, and the heat may be enough to render conditions springlike on the outer planets.

CHEMICAL ENRICHMENT

The outer layers of a giant star transport energy outward by convection in great circulating flows of gas that dip far down into its envelope. The thermal pulses set up internal convection cells that move the by-products of fusion upward. If the two convective layers can meet, freshly made elements can be brought to the top, and *the star will change the chemical composition of its surface.* Remember the stars of spectral class N from Chapter 3, the deep red giants rich in carbon? Here is their origin. At the beginning of the AGB we will see an M star whose carbon-to-oxygen (C/O) ratio is about 0.4. There is so much oxygen that it combines readily with metals to produce the characteristic TiO bands. As the carbon, the dregs of helium burning, is swept up to the top, the C/O ratio climbs. Carbon has a strong affinity for oxygen, and when C/O climbs above 1, it ties up all the oxygen in CO and CO_2. The metallic oxides disappear, and the remaining carbon combines with itself to produce the powerful C_2 bands of class N. The carbon bands are very strong at short optical wavelengths and cut out most of the blue component of the emerging light, rendering the stars a rich, beautiful red.

The deep fusion process is no longer hidden from us. We are seeing chemical elements made before our very eyes. Absolute proof of such transmutation comes from observation of the element technetium, atomic number 43. It has no stable isotopes. Even the longest-lived, ^{98}Tc, has a half-life of only 2 million years and therefore long ago disappeared from the Earth. Never buy stock in a technetium mine. But some stars, particularly those of the carbon-rich variety, show the stuff in their atmospheres! We believe it is ^{99}Tc, and it can exist only because it is being newly made and brought to the surface along with the carbon. The alchemists of old would have loved it.

The manufacture of technetium demonstrates a host of subsidiary nuclear reactions collectively called the s-process, which results from the slow capture of free neutrons onto heavy elements. Each neutron addition produces the next heavier isotope of the same element. If the isotope is stable, it eventually captures another neutron, and the mass number goes up another notch. If the nucleus is unstable, and if the capture rate is slow enough, it may beta-decay (the neutron changing into a proton with the ejection of an electron), increasing the atomic number to the next atom in the periodic table. However, before it decays, the nucleus may well catch another neutron, jumping the mass by another unit and producing the next higher isotope of the next element. In this way, an entire nuclear network of creation is established, the higher elements being made from the next

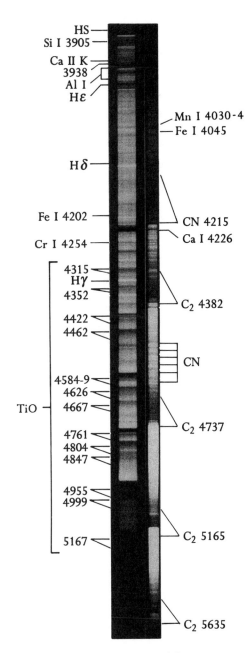

The spectrum of a normal M star (left) is dramatically different from that of a carbon star (right), whose carbon bands cut out most of the short-wave part of the spectrum.

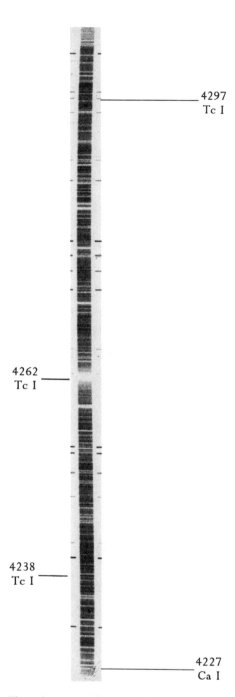

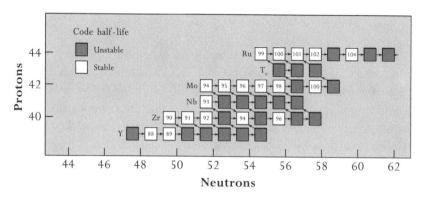

This s-process network shows how heavy elements can be made from lighter ones. The elements and isotopes are laid out in a chart of the nuclides (a nuclide is any single species of atomic isotope) in which atomic (proton) number is plotted against neutron number (atomic mass minus atomic number). Start with yttrium in its common form, ^{89}Y. It captures a neutron to become ^{90}Y and beta-decays into ^{90}Zr (the common metal zirconium). Although ^{90}Y is highly unstable, some of it will capture another neutron to become ^{91}Y, which then produces ^{91}Zr. The process works all the way to ^{94}Y. Beyond that, the isotopes of Y are violently unstable, and the isotopic upgrading stops. As the zirconium isotopes are produced, they immediately start capturing their own neutrons, so that ^{91}Zr is produced from both ^{91}Y and ^{90}Zr. But the Zr isotopes then decay into atoms of niobium, Nb. As a result, the isotopic ratios of Zr depend on neutron-capture rates relative to decay rates of the various isotopes. Molybdenum is then created from the niobium. Finally, the object of our attentions, ^{99}Tc, is manufactured by beta decay from ^{99}Mo. ^{100}Tc can be produced only by neutron capture onto ^{99}Tc because ^{100}Mo is stable. Tc turns into ruthenium, and so on.

The carbon star 19 Piscium displays strong lines of technetium, clearly showing stellar transmutation of the elements.

lower ones. The networks can be included in stellar models and isotopic ratios predicted from neutron-capture rates (which depend on temperature and density) and the decay rates of the various isotopes.

The S stars represent a transition state between the oxygen- and carbon-rich, where the abundances of the two are about equal. Titanium is much more abundant than zirconium, so that in M stars TiO predominates over its cousin ZrO. But zirconium has a much greater affinity for oxygen. Given only a few atoms of O and a mixture of zirconium and titanium, ZrO will appear first, even if the heavier metal is fairly rare. If C/O is close to 1, the C grabs all but a small part of the O, and what is left produces ZrO. Moreover, zirconium is an s-process element and is increasing in abundance along with the C. Consequently, the S stars are characterized by the ZrO bands. Not surprisingly, many of them display technetium lines as well. The

M and carbon stars both also display enrichment in nitrogen, the result of burning by the carbon cycle. Although these stars are too cool for us to see helium lines, other evidence shows that this element, too, can be enriched.

Stellar models tell us the amount of fresh material that is brought to the stellar surface and, most importantly, where in the star it came from. This information enables prediction of the strengths of various spectrum lines. The absorptions produced by different isotopes of an atom fall at nearly identical wavelengths and cannot ordinarily be separated—for example, we cannot tell one isotope of Tc from another. But *molecular* isotopic wavelength differences are relatively large. The absorptions produced by ^{12}CO and ^{13}CO or by ^{90}ZrO and ^{91}ZrO can be distinguished, allowing very detailed observational checks on the models. Although exact agreement can hardly be claimed, the results are enough in accord with theory so that we have a good deal of confidence that we know what is going on deep inside the stars as they evolve. The main problem is the convection that brings the new stuff to the surface. From observational evidence, the carbon stars seem to come mostly from bodies that started in the neighborhood of 2 or 3 $M_\odot$. However, our lack of knowledge about convection keeps us from being able to predict the exact mass ranges responsible for the carbon stars as well as accurate elemental and isotopic yields. If stellar astrophysicists had a last wish, it might be to have a theory of convection.

MASS LOSS

All stars, by their very natures, must lose mass—even the Sun, albeit at a paltry 10^{-14} $M_\odot$ per year. The rates among the giants are perhaps a million or more times greater, and on the AGB mass loss is driven at a still higher rate. The spectra of Mira stars show hydrogen emission lines that are interpreted as shock waves generated by the violence of pulsation (a shock wave is an acoustic overpressure caused by something that moves through a gas faster than the speed of sound—a supersonic aircraft provides a familiar example). The shocks, together with high stellar luminosity, loft gas off the surface into the relative cold of space. Some of the gas condenses into small solid grains, loosely called dust. Radiation pressure acting on the dust pushes it away, and the dust drags the gas along with it at a speed of a few tens of kilometers per second. The result is an outflowing dirty wind that near its maximum may reach an astonishing rate of 10^{-5} $M_\odot$ per year. The stars are in this state for more than 10^5 years and will lose substantial parts

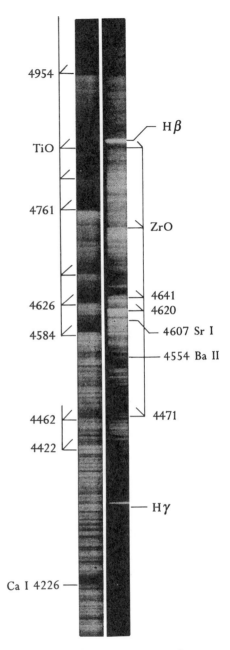

The TiO bands of the M6 giant V744 Centauri (left) contrast sharply with the ZrO bands seen in the spectrum of the S6 star T Camelopardalis (right).

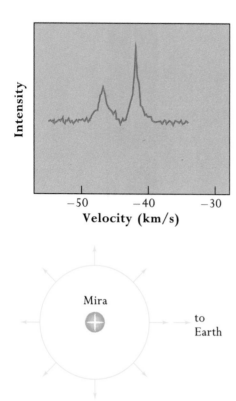

Strong microwave lines of OH, generated by a maser, are radiated from the circumstellar shell of the OH/IR star U Orionis (upper diagram). The lines are split by the Doppler shift because the cloud is expanding at a rate of about 3 km/s. The whole system is approaching Earth at a rate of about 44 km/s. The emissions are energized by the pulsating Mira (lower diagram). As a result, we see a time delay between the variation of the front side of the cloud relative to the rear side because of the finite speed of light, and can derive the size. The angular dimension then yields distance.

of themselves back into space. Between its future RGB and AGB states, the Sun will come close to cutting itself in half, and a more massive star, one that is near the upper bound of the intermediate main sequence, may lose an astounding 80 percent of its mass.

A Mira itself loses so much mass that it is surrounded by a huge expanding cloud that is visible in the infrared and radio parts of the spectrum. In the IR, the spectral signatures of dust grains around Miras reflect the composition of the underlying star. If it is carbon-rich, we will see silicon carbide grains, and if it is of the oxygen-rich M variety, silicates (silicon−oxygen compounds) appear. Radio observations are even more fascinating. The circumstellar shells of the M stars display powerful emission lines of hydroxyl, OH, and the systems are consequently called OH/IR stars. The OH is a natural maser, the microwave version of the now-familiar laser that works by pumping electrons into high-energy states. When they cascade downward, a powerful beam of light can be produced. In the lab, the laser is energized by an outside energy source such as electricity; in space, the radiation from the Mira is responsible. We also see masering lines of water and of silicon monoxide.

Part of the cloud is going away from us and part toward us. The lines are then split by the Doppler shift, and we can measure the expansion velocity. Since the OH line strengths depend on the Mira's luminosity, they will also vary. And because it takes considerable time for the starlight to reach the shell, the near side of the cloud will reflect the variations first. The time difference in variation between the near and far sides depends on the cloud's physical diameter and the speed of light. From the physical and angular sizes, we find the precious distance and can derive the stellar luminosity. These stars are treasuries of information.

The carbon stars show different but equally bizarre characteristics. Carbon forms complex organic molecules, for which conditions in space are highly favorable. In the most extreme example, a carbon star complex called IRC +10 216 (infrared carbon star, 10° declination, 21.6 hours right ascension), some 20 organic molecules are found, including such exotic ones as HC_7N and CH_3CN.

The flow of gas may be so strong that, like IRC +10 216, the star may bury itself in its own effluvia and in the optical part of the spectrum disappear from view, leaving us with only infrared and radio radiation to mark its position. As the circumstellar shell departs, it carries with it the fresh nitrogen, carbon, helium, and s-process elements that were created earlier in the star's evolution. There are thousands of these OH/IR and carbon stars. Not only are they recycling much of their matter back into space, but together they also cause a continuing large-scale enrichment of the

Not only can a photon cause an upward jump of an electron, it can stimulate an electron in an upper orbit to jump downward (left). If it does, two photons go flying away in the same direction in phase with one another. In a normal gas, absorptions and both spontaneous and stimulated emissions are in balance. If we can somehow overpopulate or pump the upper level, each photon generated produces another, resulting in a cascading beam of laser light.

interstellar gases, out of which new stars are always being formed. We now begin to see a reason for the relation between stellar metal content and age!

Mass loss may be the Earth's salvation. As solar gravity is diminished, our battered planet's orbit will slowly become larger, and the expanding Sun may not be able to reach us. Mercury and Venus are still doomed, however. Their mass may actually contribute significantly to the dust within the wind. Moreover, the great luminosity of the evolved Sun will probably melt most of the comets in the inner comet belt (just beyond Neptune), augmenting the wind's water content. It is sobering to think that the dust and the water masers we see around AGB stars may, at least in part, be a signature of the destruction of a planetary system.

The internal temperature is still going up, and we may think that at some point the carbon–oxygen mixture in the degenerate core will fire and start to burn to something heavier. But because the star has lost so much mass, the compression stops short of the required conditions. Mass loss will eventually strip away the envelope to reveal the dense dregs, the remnants that we long ago recognized as 40 Eri B and Sirius B.

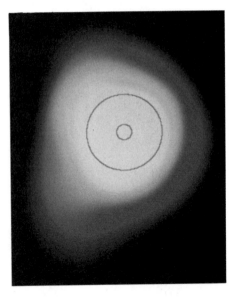

An infrared image of the carbon star IRC + 10 216 displays its circumstellar shell, which is being lost to space. Superimposed circles approximate the sizes of the star and the shell's inner edge.

PLANETARY NEBULAE

While scanning the skies 200 years ago, William Herschel came across an odd-looking body in the constellation Aquarius, a somewhat circular fuzzy patch of light now known as NGC (New General Catalogue) 7009. He called this object—and several others that he later discerned—planetary nebulae, as they reminded him of the extended disks of planets. He soon also discovered that at their centers they commonly had single stars that he realized were intimately connected with the nebulosity.

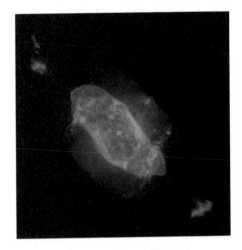

The first known planetary nebula, the Saturn Nebula in Aquarius (also called NGC 7009), wreathes a dying star.

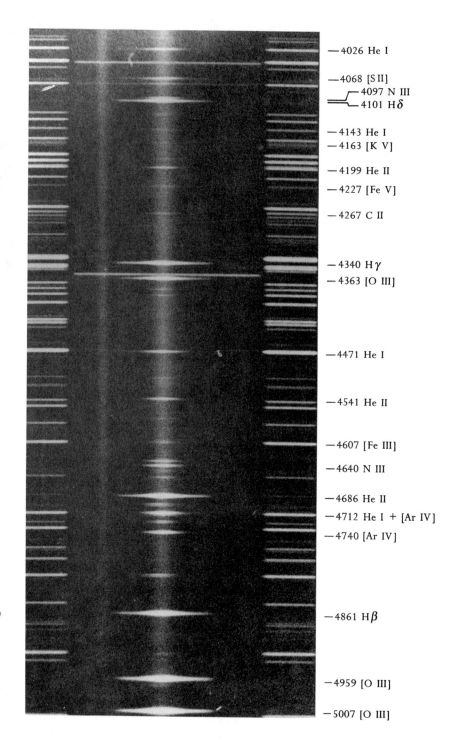

—4026 He I

—4068 [S II]
—4097 N III
—4101 Hδ

—4143 He I
—4163 [K V]

—4199 He II

—4227 [Fe V]

—4267 C II

—4340 Hγ
—4363 [O III]

—4471 He I

—4541 He II

—4607 [Fe III]

—4640 N III

—4686 He II
—4712 He I + [Ar IV]
—4740 [Ar IV]

—4861 Hβ

—4959 [O III]

—5007 [O III]

The spectrum of the planetary nebula NGC 2440 in Puppis displays a wealth of recombination and forbidden lines (the latter denoted by square brackets). The original "nebulium" lines—doubly ionized oxygen, [O III] at 5007 and 4959 Å— are seen at the bottom. Comparison spectra are seen to either side of the nebular spectrum.

The clue that led to the explanation of the planetaries was served up in 1864 by the English spectroscopist William Huggins, who discovered a trio of emission lines that proved them to be made of a thin gas. One, at a wavelength of 4861 Å, was thereafter identified with common hydrogen, but the others, at 4959 and 5007 Å, defied classification. For a time, they were thought to be produced by an as-yet-undiscovered element, which was named nebulium. After the periodic table of the elements was filled in, it was obvious that there was no place for "nebulium" and that some unknown mechanism was causing the production of the lines. As the spectrum was explored further, several more of these unidentifiable emissions were found, particularly bright ones appearing at 3868 Å, 3727 Å, and near Hα at 6584 and 6548 Å.

The solution to the mystery, and the explanation of the mechanism that lights the planetaries, was offered by Hermann Zanstra and Ira S. Bowen around 1929. Zanstra showed that the hydrogen lines were produced as a result of photoionization by ultraviolet light emanating from the central star. When an electron in the ground state of hydrogen absorbs a photon with a wavelength of 912 Å (the Lyman limit) or less, it will gain enough energy to escape the proton. A nebula is a sea of these free protons and electrons. When the two kinds of particles collide, there is a good chance that the electrons will be captured into upper orbits. Jumps to the ground state will then produce the observed emission lines. The re-formed atoms subsequently absorb more stellar photons and repeat the cycle. Helium and even oxygen, nitrogen, and neon produce these recombination lines.

Zanstra showed that in a dense nebula, every stellar ultraviolet photon will produce one ionization, which will be followed by one recapture and one observable Balmer photon (produced by an electron that lands in hydrogen's second orbit). The number of these is proportional to the ultraviolet luminosity of the star, whereas the visual magnitude is proportional to the visual luminosity. If the star is assumed to be a simple blackbody, which it generally appears to be, the ratio of the brightness of the observed nebular hydrogen emissions to the visual stellar brightness yields the effective, or Zanstra, temperature. Zanstra—who referred to this procedure of detecting otherwise invisible ultraviolet radiation as "space research at low cost"—found that the central stars of the planetaries were enormously hot. Modern work shows that they are the hottest stars known, with Zanstra temperatures that start at 30,000 K or so and extend to over 200,000 K. These are also among the most luminous stars in the Galaxy, many between 1000 and 10,000 times brighter than the Sun, but because most of their energy is radiated in the UV, we do not much notice them.

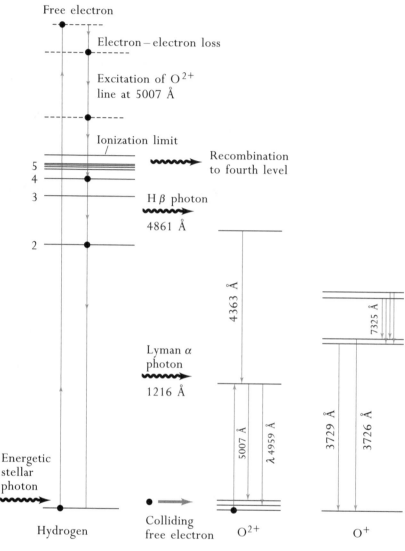

Recombination lines of hydrogen are formed when an electron is ejected (left) from an atom after the absorption of a stellar ultraviolet photon. Its energy is then greater than the ionization limit, and it flies free in the gas. It loses some energy to other electrons, in the process establishing the kinetic temperature, and then hits an O^{+2} ion, knocking its electron into an upper orbit. The free electron subsequently recombines with a proton and skips down the energy ladder to produce hydrogen emission lines. The excited oxygen electron eventually drops back down to produce one of the so-called nebulium lines at 4959 or 5007 Å. Other such lines, initially thought to be theoretically forbidden, are shown for both O^{+2} and O^+. Chemical compositions can be derived by applying basic theory and appropriate data on collision, recapture, and cascade probabilities (scales of O^{+2} and O^+ are doubled for clarity).

Bowen looked inward, to the laboratory. He used ultraviolet experiments to chart out atomic orbital structures and found that the "nebulium" lines were actually produced by energy levels of doubly ionized oxygen that lie just a bit above the ground state. By the simplest approximations of the rules of quantum mechanics, transitions between these states are not allowed, and the resulting lines are "forbidden." Only when higher-order approximations of the equations are taken into account do we see that the jumps are not truly forbidden, just unlikely. The electrons are excited into the upper levels by collisions with those electrons that were freed from hydrogen. The forbidden lines are not seen in the laboratory because the density of the gas is too high and they are weak relative to other emissions. But in the nebulae, conditions are just right, and they overpower even the recombination lines. The forbidden lines at 3727 Å, 3868 Å, and the pair in the red were subsequently found to be created by similar transitions of ionized oxygen (O^+), neon (Ne^{+2}), and nitrogen (N^+).

The theory of the production of the recombination and forbidden lines is so straightforward that we can easily use them to derive not just the central star temperatures, but also nebular densities, temperatures, and chemical compositions. We find typical densities of 100 to 10,000 atoms per cubic centimeter, comparable to the best vacua produced on Earth, and kinetic temperatures of 10,000 to 20,000 K. Compositional analyses show that these objects tend to have elevated levels of helium, nitrogen, and carbon. In a few cases, the enrichment is quite high, double the basic solar helium content, and 10 times the nitrogen.

The planetaries exhibit an almost bewildering variety of structures and are among the loveliest objects to be found in the Galaxy. Some are

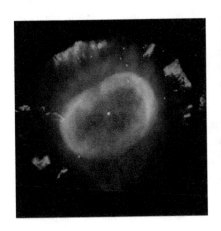 

Planetary nebulae show a wonderful variety of types. At far left, NGC 7662—the "Blue Snowball" in Andromeda—is a classic double-shell structure; its central star is still heating. Abell 39, in Hercules is shown at near left. It is huge, a parsec or so in diameter, and beginning to dissipate into interstellar space, its nucleus now a genuine, cooling white dwarf that will soon be abandoned by the expanding shell.

nearly round, others display a distinctly bipolar structure, and yet others are so complicated as to defy simple description. Many have double or even triple shells. In size they range from small and compact, nearly stellar, to dimensions so huge that they span the spaces between stars.

It was not until the second half of this century that astronomers were finally able to fit these graceful objects into the flow of stellar evolution. On both its ascents of the giant branch, an evolving star loses mass. By the time it is near the top of the AGB what is left of the star is surrounded by an enormous expanding cloud. As luminosity climbs, the mass-loss rate increases as well, so the inner part of the cloud is denser than the outer. Finally comes the big moment when nearly all the envelope is removed, not quite revealing the nuclear-burning shells. The chances are that this event happens when the helium shell is turned off, so there is now a degenerate C–O core that is successively nested in an inert helium shell, a burning hydrogen shell, and a thin though quite large hydrogen envelope about the size of the present Sun. Now, instead of coming from a distended giant, the wind must blow from a smaller star with a higher surface gravity, and it thins and increases its speed. It will eventually blow as fast as 4500 km/s. This fast wind, acting much like a snowplow, shovels the mass lost earlier into a relatively dense shell with a mass a few tenths that of the Sun. While this inner ring floats away at about 20 km/s, the hydrogen envelope of the central star thins, dissipated from above by the wind and eaten into from below by nuclear burning. The effective temperature rises at constant luminosity. When it hits about 30,000 K it produces enough ultraviolet light to ionize the surrounding mass, and suddenly a new planetary nebula blossoms into the sky.

Deep imaging will frequently show the resulting planetary buried within a vast mass of previous ejecta. The structure of a nebula apparently depends on asymmetries present in the original mass-loss flow that are magnified by the compressing fast hot wind. The composition of the object, nitrogen- or carbon-rich, tells us the kind of AGB star that produced it, giving us a direct look into the old stellar envelope and providing an excellent check on the theory of stellar structure and evolution.

The star heats to a maximum that depends on mass, 100,000 K or greater. After 10,000 years or so (depending on mass) the H envelope is nearly gone, the nuclear burning is banked if not shut down, and the star begins to cool and dim. By this time, the nebula is a tenth or two of a parsec across. Finally, after some 50,000 years, it has become so large and thin that it disappears from view. The giant has now returned a share of its matter back into interstellar space, adding gifts of helium, nitrogen, and carbon, and perhaps a load of heavier s-process elements. The next generation of

NGC 6826 is roughly circular, buried deeply within a vast outer halo produced by eons of fairly symmetrical mass loss by the parent star.

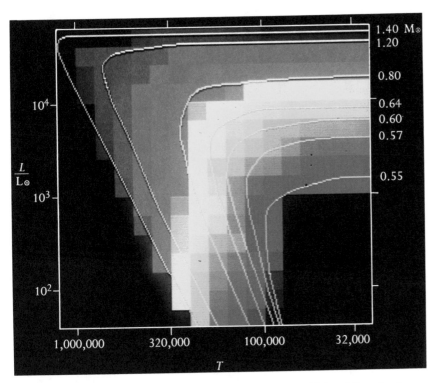

Evolutionary tracks of the central stars of planetary nebulae extend leftward from the diagram seen on page 153, their locations and maximum temperatures depending on mass. The color gives the expected relative numbers of stars at each point, with red the highest, blue the fewest. Low-mass stars linger along the horizontal (heating) portions, while those of high mass zip through this phase and linger along the cooling tracks. The observed distribution qualitatively fits this theoretical one.

stars will benefit from this largesse. What remains is a hot, naked, white dwarf. There is evidence that not all of the white dwarfs go through this stage, but it is clear that the majority do. Those that do not somehow manage to pass directly from the horizontal branch; we do not know how.

WHITE DWARFS AT LAST

As the end products of the intermediate main sequence, white dwarfs are everywhere. Huge numbers roam free in the Galaxy, as will the degenerate descendent of the Sun. They do not obviously populate the sky only be-

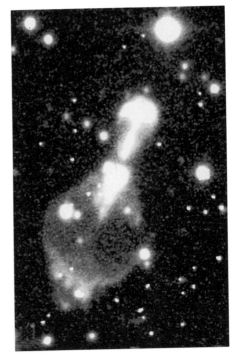

The Calabash Nebula displays huge bipolar bubbles, produced by the wind of a Mira buried at the center. They seem to have been focused by a disk surrounding the star that allows matter to pass outward in opposite directions. Astronomers of the distant future will probably see an extreme bipolar planetary develop in the center.

cause they are so faint; even the brightest is only eighth magnitude. It took billions of years to make the AGB star—then in a blink it is gone, replaced by its defunct, degenerate core. Now nothing remains for it but to cool forever, to become a clinker in the celestial ash heap.

Although the huge densities of a million grams per cubic centimeter make these stars strange enough, there are more surprises. When the spectrum of Sirius B was first observed, the star was classified as type A because of the great strengths of its hydrogen lines. Sirius B is hot, however, and actually has a temperature that would place it in class B, which ordinarily exhibits helium lines, too—yet none at all are found. Other white dwarfs that do display helium absorptions were called class B. All real B stars have

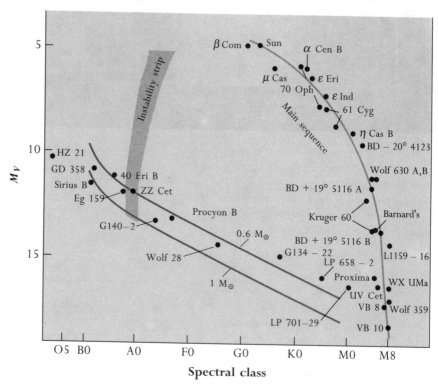

The white dwarfs are seen cooling along their final evolutionary tracks in this detail from the HR diagram shown on page 89. Theoretical tracks for 0.6 $M_\odot$ and 1.0 $M_\odot$ have been superimposed. White dwarfs are not classifiable by ordinary rules, so they have been placed here according to their temperatures and the temperature scale of the main sequence. The more massive white dwarfs are less luminous, because higher gravity gives them smaller surface areas; at the lower end the stars will crystallize, but the lifetime of the Galaxy is not yet long enough for any of them to have cooled to invisibility.

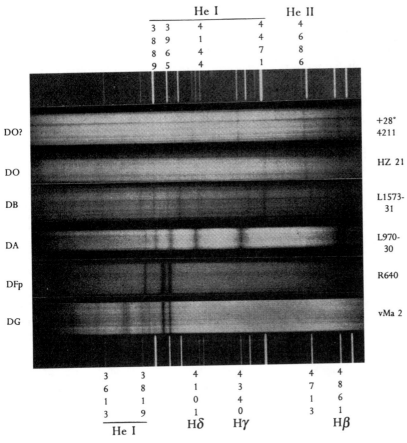

White dwarf spectra show a strange selection of absorptions. They were originally classified, as were all other stars, by the kinds of lines in their spectra, but the DO and DB stars at the top have only helium lines, whereas DA has only hydrogen. In the stars below, neither element is visible, and we see ionized calcium. These degenerates are chemically differentiated. The helium- and hydrogen-rich stars, now called only DB and DA respectively, pervade the whole temperature sequence.

hydrogen lines, but these did not. The white dwarfs are chemically differentiated into two broad groups. The hydrogen- and helium-rich degenerates are respectively called DA and DB after their initial classifications, but there is little relation to temperature: Each kind spreads out across the whole temperature sequence. Even though the cooler white dwarfs can exhibit no lines of either element because of their low temperatures, we are confident that the difference extends to them as well.

If you could stand on the Sun (an interesting thought), you would weigh triple what you do on Earth, the result of a 300,000-fold increase in mass and a 100-fold increase in radius. On a white dwarf, however, the radius is so tiny that a typical 70-kg human would weigh an astonishing 600,000 kg! The force of gravity is so high that the heavier atoms, including those of helium, are dragged to lower depths in the atmosphere and the hydrogen floats to the top. The result is a pure hydrogen gas and a DA star, no matter what the temperature. Given this diffusion process, we know that a helium-rich DB star can contain no hydrogen at all, or it would be seen. It must have lost its entire hydrogen envelope during evolution out of the AGB state. We can actually see this differentiation start with the planetary central stars, but we do not know why one star loses its hydrogen skin and another does not. Stranger still, there is a gap between 45,000 K and 30,000 K in which there are no DB stars. Apparently, the degenerates can change from one type to another as they cool! Helium may be lofted back to the surface of a DA, the hydrogen of a DA may be lost through a weak wind, or a DB may accumulate hydrogen from the interstellar medium. But we really do not understand.

The magnetic fields possessed by some of the stars are more remarkable yet. The Zeeman effect causes the absorptions of the magnetic A dwarfs, which have fields up to a few thousand times that of the Earth, to be split by a few tenths of an Ångstrom. In the class of magnetic white dwarfs, the splitting can be so huge that the lines become unrecognizable, Hα and Hβ spread out over much of the visible spectrum. To produce these prodigious separations, fields of the order of *hundreds of millions* of times that of Earth are required. Somehow, nature squeezes the stellar field along with the star to produce such intense magnetic concentrations. Again, we do not know why or how, nor why some white dwarfs are magnetic and others show no fields at all.

Had enough weirdness? Allow one more. As the stars cool to effective temperatures of roughly 4000 K they start to *crystallize!* We cannot see this development actually take place because the atmosphere is still a gas, but the theory seems to make it necessary. The cooling time of a white dwarf is

The Hα and Hβ lines of the magnetic dwarf PG 1533-057 are grossly split by a powerful magnetic field 60 million times greater than Earth's.

enormously long, and it takes billions of years to descend to this point on the HR diagram. The time required for a star to move *off* the diagram is longer than the age of the Galaxy. No white dwarf ever born has ever disappeared. They are all still there and visible to us. In principle, we can use the end of the white dwarf sequence, near 4000 K, to date the Galaxy. Current work gives a lower limit of about 10 billion years, consistent with the ages derived by other means.

The most massive white dwarf known is only somewhat heavier than the Sun, reflecting the profound distinction between the upper and intermediate main sequences mentioned earlier. Now it is time to look at the reason. In 1930, a future Nobel-prize–winning physicist named Subrahmanyan Chandrasekhar traveled from his native India to Cambridge University to begin graduate studies. During the voyage he explored the theory of the internal structures of the white dwarfs and degenerate matter. He realized that under extreme conditions, the velocities of the most energetic electrons could approach the speed of light. It was necessary then to take into account the general theory of relativity. When he did, he discovered that the laws that govern degeneracy change. If the internal pressure could be elevated enough, degenerate electrons could no longer keep a star buoyed up. It would collapse if the degenerate mass could exceed what is now known to be 1.4 $M_\odot$. Above this Chandrasekhar limit, no white dwarf can exist.

As initial mass increases upward along the main sequence, so must the mass of the core. The dividing line between the upper and intermediate main sequence occurs at an initial mass of about 8 $M_\odot$, where the core reaches the Chandrasekhar limit. Above 8 solar masses, some quite wonderful things happen, but these are saved for Chapter 6.

Subrahmanyan Chandrasekhar (1910–1995).

BINARIES

The white dwarfs hold yet another surprise that renders them wonderfully visible. The majority of stars reside in binary systems, and because the masses of the two partners are usually different, some interesting things can happen over the course of evolution. Look first at our old friend Sirius. The white dwarf B component has less than half the mass of the brilliant primary. Since it had to evolve first, it must once have been the more massive, and clearly has lost over half its initial matter, providing direct proof of mass loss. Sirius A is something of an Am star, mildly enriched in metals.

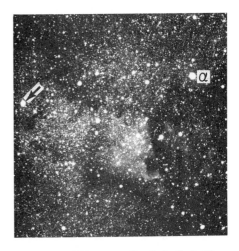

Nova Cygni 1975, next to Deneb (α Cyg). We see here an explosion that took place on the surface of a white dwarf companion to a main sequence star.

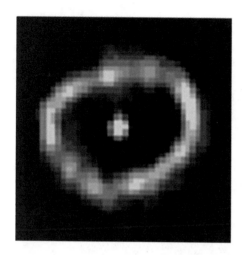

Just over a year after the explosion, Hubble observed a cloud expanding around Nova Cygni 1992. The star in the center is the ordinary dwarf that is still passing matter onto a white dwarf too faint to be detected.

We believe that when Sirius B was a giant, it passed some of its enriched surface material onto Sirius A through its outflowing wind. A whole class of giant oddballs called barium stars seem to have been made in this way.

But this kind of mild interaction is nothing compared to what else can happen. In 1975, a brilliant "new star," or nova (from the Latin for "new"), flared up in the constellation Cygnus to rival first-magnitude Deneb. Dozens of these celestial explosions are seen every year, though most are so far away as to be invisible to the naked eye. The process begins with main sequence companions. The more massive component starts to evolve. Unlike the Sirius pair, the two are so close that the envelope of the giant actually encompasses its neighbor. Orbital energy will be lost, and the pair will spiral closer together. By the time the first star finishes its evolution, the two are so intimate that the white dwarf raises tides in the remaining main sequence star, tides so great that the ordinary dwarf fills its gravitational equilibrium surface and starts to send fresh hydrogen cascading down onto the degenerate. The hydrogen does not flow directly to the companion star, but first accumulates into an orbiting accretion disk from which it is transferred to the surface. The layer of new hydrogen builds up until the temperatures become so high that fusion is initiated, the surface explodes, and the nova (a terrible misnomer, as the event is caused by an *old* star) lights the sky. It disappears after a few weeks or months, and after some years we can photograph the expanding cloud. The process repeats itself, and our descendants a hundred thousand years hence may see another.

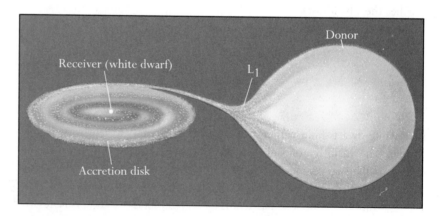

One member of a binary fills an equilibrium surface over which the gravitational strengths from the two are equal. The large one can then pass matter to the small one via an accretion disk through point L_1.

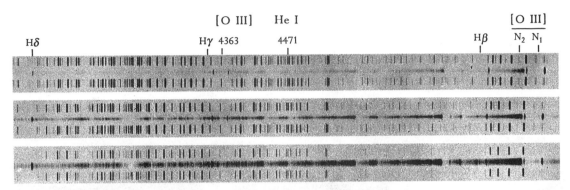

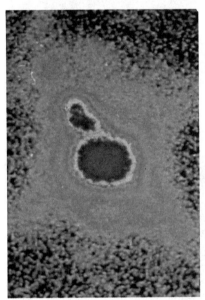 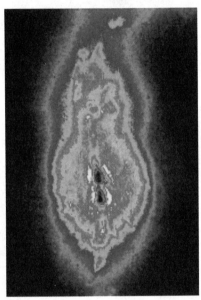

R Aquarii is a symbiotic Mira, a giant passing matter onto a white dwarf. Vigorous mass loss produces a complex nebula. At the far left, a photograph shows a jet emanating from the central star; (near left) a high resolution image of the core itself made with the Hubble Space Telescope. Spectra of the star made in the 1930s (above) show considerable variation and display emission lines superimposed on an M-star background.

Yet another scene: The giant companion of a white dwarf expands and throws copious matter into an accretion disk. The disk, or a spot on the star where the matter actually falls, is so hot that it produces considerable ultraviolet radiation, enough to ionize the surrounding matter. A spectrogram of the star then shows a weird mixture of TiO bands on a red continuum, emission lines of H, He, and even some forbidden features. This symbiotic star may even occasionally produce novalike detonations and brighten considerably, burning for years. A few generate wonderful surrounding nebulae reminiscent of the planetaries, but much more complex.

So concludes the picture of the evolution of these stars. We begin here to see how the Galaxy and the stars are inextricably locked together, each one feeding the other. More wonders await in the next chapter, where we will examine the amazing high-mass stars.

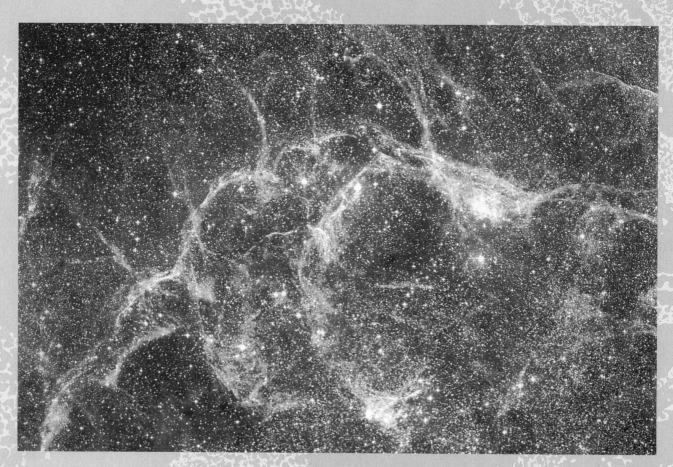

The Gum Nebula in Vela is one of only four supernova remnants for which the collapsed stellar remains can be identified.

Chapter Six

CATASTROPHE

THE SPECTACULAR LIVES AND DEATHS OF THE HIGH-MASS STARS

Some January evening, step outside to admire Orion, his belt riding the celestial equator. From it hangs his mighty sword. Focus on the middle star, θ^1. Raise a pair of binoculars and discover that it is quite fuzzy; a small telescope easily shows an intricate cloud of gas, the Orion Nebula, the archetype of the gaseous or diffuse nebula. In the center is a quartet of stars, a gravitationally bound group called the Trapezium. One, θ^1 Ori C, is notably brighter than the others. Its spectrum shows it is a hot main sequence O6 star with a temperature near 40,000 K. Its luminosity is a quarter-million times that of the Sun, and it produces so much ultraviolet radiation that it is responsible for ionizing a cloud with a mass of several hundred Suns out to a distance of nearly 4 parsecs.

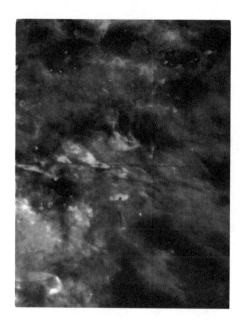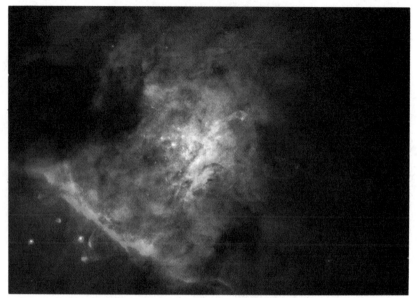

A Hubble image reveals the graceful sweeps of ionized gas that characterize the Orion Nebula (right), 420 pc away and 7 pc across. At the center is the Trapezium, the brightest of whose four stars, θ^1 Ori C (O6 V), is largely responsible for the ionization of the object. A zoom in on a portion of it (left) shows complex filamentary structures.

Orion and its great nebula exemplify the properties of the rare stars of the upper main sequence. They are enormously luminous. Their effective temperatures must be greater than 25,000 K—spectral classes hotter than B2—in order to produce enough ultraviolet photons to ionize a surrounding nebula. They cling not only to diffuse nebulae, but also to the OB associations that dot the thin galactic plane, a pattern that testifies to their youth. The associations are observed to be expanding and disintegrating, and the nebulae are the remnant clouds of interstellar gas out of which the stars were born. Even moving at several tens of kilometers per second, the O and B stars never get very far away from their origins before they begin to die. Spectral class B2 corresponds to a 10-solar-mass star, close to the nominal lower limit for the upper main sequence. The main sequence stars that illuminate the gaseous nebulae are not expected to produce white dwarfs. What they *will* produce is the main subject of this chapter.

THE DIFFUSE NEBULAE

Since the most obvious manifestations of the great O stars are the diffuse nebulae, it is appropriate to look at these wonders in more detail and to make a small survey of the brighter ones. Many are easily observable with only a small telescope or even binoculars. Half a dozen even made it into the famous list of bright nonstellar objects compiled by the eighteenth-century comet-hunter Charles Messier; the Orion Nebula is M 42.

Stars are descended from the gas and dust clouds of interstellar space. The less massive ones have little effect on their parent clouds and live so long that they escape to roam free in space. The birth of an O star, however, is marked by a vast zone of ionized hydrogen gas, the H II region or diffuse nebula. The Orion Nebula is actually a small blister on the edge of a giant cold, neutral cloud that has a mass 100,000 times solar and occupies most of the southern part of the constellation. These bubbles of ionization are called Strömgren spheres after the Swedish-born American astrophysicist Bengt Strömgren (1908–1987). The star maintains high ionization to the point at which the ultraviolet photons are exhausted, whereupon the cloud suddenly becomes neutral. One of the best examples of such a structure is in Orion's eastern neighbor Monoceros, a beautifully symmetrical billow called the Rosette Nebula. Within it is a dense cluster of stars, and both nebula and cluster are centered in the huge Monoceros OB2 association (the second OB association in Monoceros). Both it and the Orion Nebula show superimposed dark clouds that contain dust grains mixed within the gas, the same kind of dust that gives the Milky Way its structure.

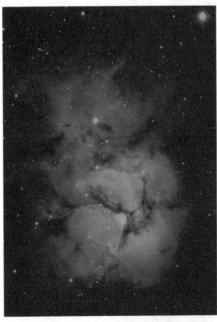

The colors of summer's lovely Trifid Nebula, M 20 (above), differentiate an H II region (red) from a reflection nebula (blue). The Rosette Nebula in Monoceros (left) is a fine example of a Strömgren sphere. Superimposed upon the apparent surface of the nebula are large numbers of black spots (globules) and long dark strings ("elephant trunks") —actually foreground clouds of opaque dust.

The northern summer sky, which contains the thicker part of the Milky Way, is richer ground. In Sagittarius is the most visible of all diffuse nebulae, the Lagoon, M 8. Just above it is the delightful Trifid Nebula, M 20. Better than any other, it illustrates the effect of stellar temperature and the division between the upper and intermediate zones of the main sequence. The top part glows bright red, the color produced by the Hα recombination line, showing that the gas is ionized and that the illuminating star must be hot and massive. The lower part contrasts smartly in blue. This part of the nebula is illuminated by a cooler (but still blue) B star. It throws out insufficient ionizing photons, and therefore the gas remains neutral. The blue starlight, however, is scattered by dust grains that are always inter-mixed with the gas to produce a reflection nebula. Proceeding northward into Cygnus we find H II regions that are more spread out. Near Deneb is the North America Nebula, NGC 7000, which can actually be outlined on a dark night with the naked eye.

The southern hemisphere has even richer treasures, most notably the Eta Carinae Nebula. This huge complex is lit by enormously massive stars and contains both the weird supergiant after which it is named and the great luminary of the Galaxy itself, HD 93129A. For the greatest of all H II regions, however, we have to leave our own Galaxy and go to the Large Magellanic Cloud. There, off to one side, is the Tarantula Nebula, so immensely bright that it actually received a Flamsteed number, 30 Doradus. This object, called a giant H II region, is so large that if placed at Orion Nebula's location 420 pc away, *it would fill the entire constellation.* For reasons we do not really understand, giant H II regions seem to inhabit the smaller galaxies and the spirals that have wide-open arms. Our own Milky Way has none. They are so large that they can be illuminated only by huge crowds—clusters—of O and B stars. Some of the most massive stars known are in 30 Dor, brilliant O stars that will one day evolve with devastating effect.

The bright Eta Carinae Nebula (top) dominates its section of the Milky Way. It is lit by several massive stars, including the most luminous star known in the Galaxy, HD 93129A. The Large Magellanic Cloud (bottom) shows the Tarantula Nebula is at top center.

SUPERGIANTS

Before it reaches its finale, however, an O star evolves first into a supergiant. The name implies that these stars are just extreme examples of the giants; they are not. True, there is a continuum in empirical luminosity classification from the subgiants through the red giants to the red supergiants—just look at any of the HR diagrams. In an evolutionary context, though, the red supergiants are profoundly different. At their peaks, the more massive of the AGB stars may reach absolute bolometric luminosities that are actually greater than those of the lesser supergiants, and even by visual magnitude they can be rivals. Yet they remain giants because they will eventually produce white dwarfs. The supergiants, which have evolved from high-mass stars, above 8 or 10 solar masses, will not.

The relation between the upper main sequence and the supergiants is evident from stellar distributions. The giants, progeny of the intermediate main sequence, are found everywhere, even far out into the galactic halo. Supergiants, on the other hand, are spread thinly along the plane of the Milky Way just like the O and B stars. Furthermore, they inhabit the same OB associations; red Antares is part of Scorpius OB2 and Mu Cephei, a member of Cepheus OB2. Since the red supergiants are still close to their places of origin, the time scale of evolution to the red supergiant state must be very short, measured only in millions of years, again confirming evolutionary theory. Finally, we can find red supergiants coupled to class B companions. Since to have evolved first they must initially have been the more massive of the pair, they have to be descendants of B or even O dwarfs.

The naked-eye red supergiants, found along the galactic plane in the Milky Way, have the same distribution as the O and B stars, the result of a direct evolutionary relationship.

The internal evolution that develops the supergiants is at first similar to that leading to the giants. As hydrogen is consumed in the deep cores of these stars, the points that represent them on the plane of the HR diagram move up and to the right, giving considerable breadth to the main sequence. By the time the initial fuel has disappeared, most of them have changed into B1 or B2 dwarfs, and their effective temperatures have cooled to the point where they can no longer ionize any surrounding gas. Diffuse nebulae, if any are present, will shortly turn into reflection nebulae. As the now-quiet helium cores contract under gravity's squeeze, the stars depart the main sequence and cool at nearly constant luminosity. They first become blue supergiants like Orion's Rigel and then pass into type A, to appear like Deneb.

From about 40 solar masses down, the stars continue cooling all the way through class K. As they enter M, the core temperatures reach 150 million degrees, and the helium will again fire up and burn via the 3-α process into carbon and then oxygen. The stars have now arrived in the realm of the classic red supergiants and are monsters like Antares, Betelgeuse, and Mu Cephei. Yet the red supergiants never get as cool as the

Evolutionary tracks of massive stars show that they move at about constant luminosity, to the right where they become supergiants. From about 40 $M_\odot$ down, they pass all the way from class O to class M as the interiors contract and then begin to burn helium into carbon. As the stars below 12 $M_\odot$ move back and forth across the instability strip (cross-hatched zone), they become high-luminosity Cepheids. Above about 60 $M_\odot$ the stars do not make it to class M, but fire their helium in hotter spectral classes. All the time they are shedding huge amounts of mass. The shaded areas show regions of stable core burning where stars concentrate. To the left, they burn hydrogen on the main sequence; to the right, helium.

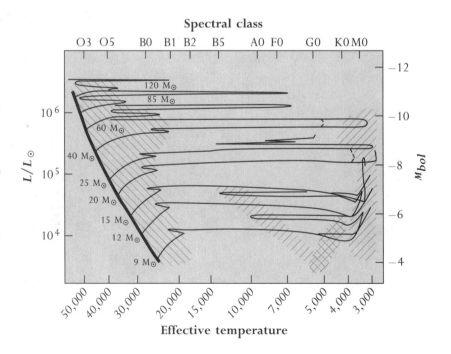

advanced AGB stars, another difference between the two kinds. Because of rapid evolution through the middle part of the diagram, F and G supergiants are rare. As a result, these massive evolved stars divide into two broad classes, cool red and hot blue. The ratio of the two groups is an important test of astrophysical theory, which passes. Below about 12 M$_\odot$ the evolutionary tracks on the HR diagram cross the instability strip and the stars become the most luminous Cepheids, first during the transit from blue to red and then as their tracks loop back and forth.

Above 40 M$_\odot$ the evolutionary pattern changes, largely as a result of extreme mass loss. The stars do not become red, but instead fire their helium within class A or the hot end of F. There are few to start with and their evolution is rapid, so there are not many of them to admire. Finally, whatever the mass might be, the helium runs out, resulting in a star with a carbon–oxygen core surrounded by burning shells of helium and hydrogen. The speed of the process is astonishing. A 20-M$_\odot$ star will complete hydrogen burning in only 8 million years, and helium burning in a mere 1 million. A million years may seem like forever to us, but remember that the Sun takes 10 *billion* years just to complete hydrogen burning.

After this point, the internal evolution of upper and intermediate main sequence stars is totally different. The AGB stars quit burning at their centers when they develop degenerate carbon–oxygen cores. Their masses are too low to allow the interiors to get hot enough to fuse the carbon into other elements. Core contraction then stops as a result of degeneracy pressure. In contrast, the supergiants are so massive that the burning interiors are beyond the Chandrasekhar limit. However, their higher temperatures and lower densities keep them from becoming degenerate and collapsing. When their contracting carbon cores reach an incredible billion degrees, the carbon begins to fuse, creating another stabilizing energy source and a growing mixture of neon, magnesium, and oxygen. From around 20 solar masses down, C burning takes place among the red supergiants. It is impossible to identify specific stellar candidates. The phase takes such a short period of time that their number must be very small. In the upper reaches of mass, above about 40 M$_\odot$, the stellar evolutionary tracks loop all the way back across the HR diagram, and the stars are reborn as blue supergiants before carbon is ignited.

The end of evolution is not yet at hand. The Ne-Mg-O mixture will eventually burn into something else, the fusion machine creating heavier elements until it is stopped by sudden death. But there are yet more acts before the curtain is lowered. The supergiants are so luminous that they lose mass at a prodigious rate, their extreme winds producing some of the stranger celestial performers.

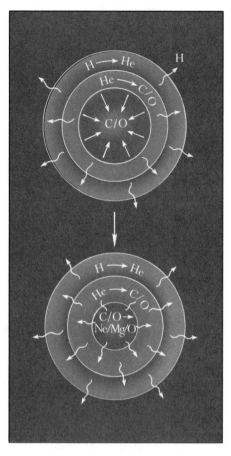

The evolution of a supergiant differs from that of a giant. The carbon–oxygen core contracts and the carbon eventually begins to burn to a mixture of neon and magnesium and oxygen. The entire nuclear-burning region, which includes shells of fusing helium and hydrogen, is well past the Chandrasekhar limit, and is not degenerate. No white dwarf can form.

Mira, the prototype of a giant long-period variable, displays emission lines of hydrogen in its spectrum and is therefore known as an M7 IIIe star. The pioneers of spectral classification gave the O stars last place in the stellar alphabet because many of them were also seen to have emission lines. It was not until well after tens of thousands of stars had been typed that Antonia Maury realized on the basis of continuity of the absorption lines that the O stars should be placed at the beginning of the sequence, not at the end. The O stars display such a variety of emissions that the simple letter "e" does not do justice to them. Some have emissions of hydrogen, others display nitrogen and helium; the latter have f rather than e appended to their spectral classes. The e and f suffixes signal powerful mass loss, a unifying characteristic of supergiants.

For the most massive stars, the Oe and Of states are the first observational stages following the main sequence. As they brighten, these stars begin losing mass and become surrounded by circumstellar shells that produce the emissions. Matter is being stripped away. At the same time, some of the by-products of thermonuclear burning can make their way to the surface to enhance the nitrogen lines that give rise to the Of class. The most intrinsically luminous star known, HD 93129A, which is involved in the Eta Carinae Nebula, is an O3 If supergiant.

In his *Uranometria* (1603), Johannes Bayer named a relatively obscure star in Cygnus "P"; it is one of the few that still retain their old roman-letter designations. The spectrum of this star is outlandish, showing weird emission lines that are flanked to the violet by absorptions. These features are the special signatures of dense winds. The emissions again show that a hot, low-density gas surrounds the star. But along the line of sight the wind is superimposed on the stellar surface, and it is dense enough to create deep absorptions that are Doppler-shifted to shorter wavelengths. All emission–absorption combinations of this kind have become known as P Cygni lines.

Emission lines of nitrogen and helium make HD 190429 both an O4 supergiant and an Of star. Not only has it started losing significant mass, but by-products of the carbon cycle seem to have worked their way to the surface.

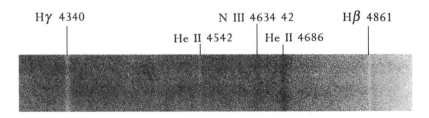

Hγ 4340   N III 4634 42   Hβ 4861
He II 4542   He II 4686

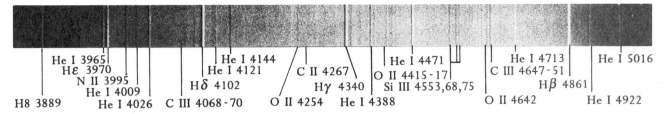

He I 3965	He I 4144		He I 4471	He I 4713	He I 5016
Hε 3970	He I 4121	C II 4267	O II 4415-17	C III 4647-51	
N II 3995			Si III 4553,68,75		
He I 4009	Hδ 4102	Hγ 4340		Hβ 4861	
H8 3889	He I 4026 C III 4068-70	O II 4254 He I 4388	O II 4642		He I 4922

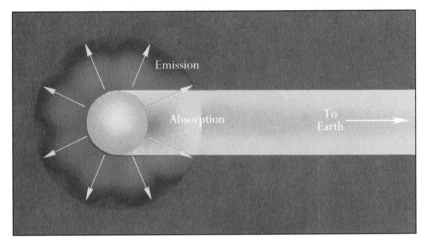

Emission

Absorption

To Earth

The weird spectrum of P Cygni has emission lines coupled to absorptions on their violet edges. Such lines are formed by a rapidly expanding cloud of gas that results from mass loss. Most of the low-density cloud will produce emissions; however, some of the matter is flowing directly at the Earth with the star in the background, and will instead create absorptions. The high velocity of the gas will Doppler-shift the absorptions to shorter wavelengths, producing emission lines flanked on the blue side by absorption lines.

They are a fount of information. The point of minimum wavelength at which we see absorption corresponds to the matter that is coming directly at us, allowing the determination of the wind's velocity. From the strengths of the absorptions and emissions, the amount of matter in the wind can be found, which when combined with the velocity yields the mass-loss rate. The Of stars, those near the beginning of their evolution, display P Cygni features in the deep ultraviolet parts of their spectra. Here we find spectrum lines that arise from the lowest orbits, the ground states, in which most of the atoms reside. At these wavelengths the gas is sufficiently opaque to form the absorption components. The extreme Doppler shifts of the absorptions indicate winds that can blow in excess of 4000 km/s. In extreme cases, the stars create visible bubbles and have powerful impacts on their surroundings.

P Cygni itself, a B1 Iae supergiant ("Ia" for the brightest of the supergiants, "e" for hydrogen emission), is one of the brightest stars in the Galaxy, with an absolute bolometric magnitude approaching −11. It is an extreme example of a windy star. There is so much matter surrounding it that even the optical lines, those that arise from the less populated higher

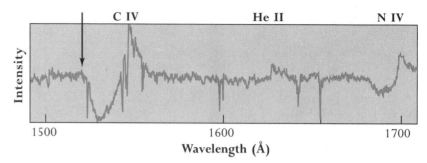

C IV   **He II**   **N IV**

Intensity

1500   1600   1700

Wavelength (Å)

Brilliant southern Zeta Puppis is an O4 If supergiant. A spectrogram taken with the International Ultraviolet Explorer displays a powerful P Cygni line of triply ionized carbon (C IV) at 1550 Å. The arrow shows the shortward limit of the absorption component, which must correspond to the fastest moving matter, that coming right at us. From its Doppler shift, we find a velocity of over 4000 km/s. A weaker feature of N IV is seen at right.

orbits, can form absorption components. It is losing matter at the astonishing rate of 10^{-4} solar masses per year. Since Homeric bards composed the *Iliad* the star has pumped a third of a solar mass into space! And it does not do so quietly. In 1600, before P Cygni acquired its name, it had blossomed to first magnitude and was recorded as a nova, which it definitely is not.

It appears to be related to an even odder star, Eta Carinae, after which the vast southern nebula was named. Eta itself is now visually obscure, of the sixth magnitude and only barely visible to the naked eye. But

The radiation from the Of star HD 148937 lights up a huge complex of gas and dust in the Ara OB1 association (left). Immediately surrounding the star is the nebula NGC 6164-5 (right)—recent ejecta formed by the star's fierce wind.

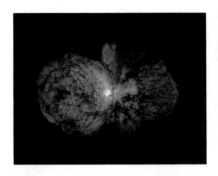
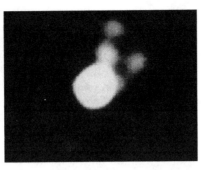

in 1844 it was brilliant, second only to Sirius. Then it began to fade, and only a dozen years later had dropped out of sight to seventh magnitude. It has only brightened a bit since, and even more remarkable, its spectrum has changed too. In the 1890s it appeared as an F supergiant. Now in the optical part of the spectrum it displays neither absorptions *nor* emissions. Instead it is hidden in a dusty nebula called the homunculus that radiates its own emission lines.

But a star of class F cannot possibly ionize a nebula. We believe that η Car actually contains a masquerading B supergiant that has lost so much mass that it has simply buried itself in its own effluvia. The class F spectrum was not that of the star, but of the thickening surrounding shell. Filaments of gas around the bright nebula show directly that this kind of activity has been going on for a thousand years or more. One hundred fifty years ago, Eta Carinae suddenly erupted, producing an extraordinary burst of matter in a wind outflow as high as a tenth of a solar mass per year in which as much as a solar mass may have been lost. Today we see the result of the event as an immense bipolar flow set perpendicular to a thick central waistband. The culprit itself is among the most massive stars in the Galaxy, weighing in at 100 suns.

Eta Carinae seems to be an example of one of the rarest breeds of stars, the luminous blue variables (LBVs). P Cygni is another, and so may the star Rho Cassiopeiae, which in 1945 dropped from magnitude 4.5 to 6 as its spectrum wandered from F to M; it even developed TiO bands! Again, the phenomenon responsible was probably a dusty shell caused by a mass outflow rate of 10^{-4} solar masses a year. These stars are so bright that they can easily be seen over intergalactic distances. Four of them, originally called Hubble—Sandage variables after their discoverers, are seen in the nearby spiral galaxy M 33. Like Rho Cassiopeiae, one of them recently dropped 3.5 magnitudes as its spectrum changed from F to M. The small number in a

NGC 2359 in Canis Major is a dense shell surrounding the nitrogen-rich Wolf–Rayet star HD 56925.

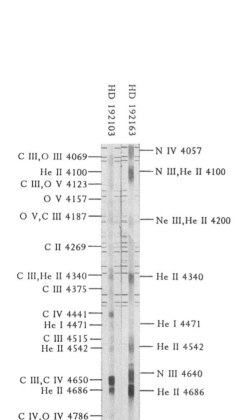

C III,O III 4069 —
He II 4100 —
C III,O V 4123 —
O V 4157 —
O V,C III 4187 —
C II 4269 —
C III,He II 4340 —
C III 4375 —
C IV 4441 —
He I 4471 —
C III 4515 —
He II 4542 —
C III,C IV 4650 —
He II 4686 —
C IV,O IV 4786 —
He II 4861 —

— N IV 4057
— N III,He II 4100
— Ne III,He II 4200
— He II 4340
— He I 4471
— He II 4542
— N III 4640
— He II 4686
— He II 4861

HD 192103
HD 192163

WC8
WN6

Wolf–Rayet WN and WC stars (right and left, respectively) display powerful broad lines of helium and of nitrogen and/or carbon. There is no evidence for significant hydrogen, implying that the stars have stripped themselves nearly to their nuclear-burning cores.

whole galaxy reflects the rarity of the stars at the top of the main sequence. We have no idea what causes such erratic changes in mass-loss rates.

No less strange are the Wolf–Rayet stars. These show only broad emission lines and no trace of the most common of all elements, hydrogen, although helium abounds. Mass loss seems to have been so vigorous that the stars *have lost their entire hydrogen envelopes.* They come in two very distinct varieties, carbon-rich (WC) and nitrogen-rich (WN). The WN stars have a normal complement of carbon and oxygen, but the WCs have no nitrogen at all. The ionization level of the Wolf–Rayet stars is very high; lines of He^+, C^{+3}, and N^{+3} dominate. The temperatures required to produce such ions relate these stars strongly to class O, although the emissions are so powerful that we are really unaware of the stellar surface, if one actually exists. The strong broad lines indicate current mass-loss rates of 10^{-5} to 10^{-4} $M_\odot$ per year.

Several Wolf–Rayet stars are members of binary systems, and one, γ Velorum, is even visible to the naked eye. From orbital analysis, we find that their masses are high, averaging about 20 $M_\odot$, but nowhere near the mass limit. Their companions, however, are invariably O stars that are closer to the top, from which we deduce that the W–R stars have lost at least 40 percent of their original matter, consistent with the current mass-outflow rate. The stars have lost so much of themselves and are so hot and luminous that many are surrounded by extraordinary gaseous nebulae of their own making, massive analogues to planetary nebulae.

We think that the two varieties may be related to different modes of nuclear fusion. The high nitrogen in the WN stars is a by-product of the carbon cycle, which explains why carbon is also present. On the other hand, the carbon-rich variety may be the result of the next stage, helium burning. If that is the case, then the WC should succeed the WN; but the masses of the two average about the same, so we cannot prove on that basis that evolution has taken place from one kind to the other.

The blue main sequence and supergiant stars have higher maximum luminosities than do those on the red side. There seems to be a distinct upper envelope on the HR diagram, called the Humphreys–Davidson limit, that slopes downward toward lower temperatures among the O and B stars and then levels off at about B3. The flat portion corresponds to stars of

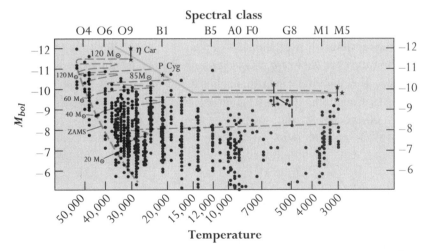

The observed separation of red and blue supergiants is obvious; the maximum brightness of the blue stars is greater. The solid yellow line shows the empirical upper boundary, the Humphreys–Davidson limit. The evolutionary tracks from the diagram on page 178 above have been superimposed as dashed lines. Only those under about 40 $M_\odot$ make it all the way across; those of higher-mass stars hit the limit. The star symbols show luminous blue variables and other stars with high rates of mass loss that may be turning into Wolf–Rayet stars.

about 40 solar masses. Below that limit, stars can evolve horizontally all the way across the diagram, changing from blue supergiant to red. Much above 40 $M_\odot$, however, the enormous luminosities produce such high mass-loss rates that the stars' redward evolutionary progressions on the HR diagram stall within class B, and they are prevented from becoming red supergiants. The LBVs and other stars with extensive surrounding shells lie on the Humphreys–Davidson limit, P Cygni and probably η Carinae right on the high-mass descending slope. They are now stripping away their hydrogen envelopes, and before our eyes may be turning themselves into Wolf–Rayet stars. In the distant future astronomers may see the most massive component of Eta Carinae as such a star. It will be accompanied by its less massive class O companions, which will not yet have had time to evolve, much as we see so many WR stars today.

SUPERNOVAE

Wolfgang Schuler, a resident of Wittenberg, must have been amazed the night of November 6, 1572 (by the Julian calendar) when he looked up to find a new resident shining among the stars of Cassiopeia. Five days later that eminent surveyor of the skies, Tycho Brahe, saw the same object for the first time, his view of it almost certainly delayed by obscuring Baltic Sea clouds. The star overwhelmed its stellar neighbors, at its peak reaching apparent magnitude -4, rivaling Venus, and visible in full daylight. Though many people saw it, only one thought to record its progress, and it is known in commemoration as Tycho's Star. After 100 days it was still as bright as Vega, magnitude zero, and it did not completely fade from sight until March 1574. Ordinary novae are not all that uncommon; in this century alone eight have attained first or second magnitude. However, the brilliance of Tycho's set it apart. Another, which appeared in October 1604 in Ophiuchus, was similarly striking—this one described by the great Kepler, Tycho's one-time assistant. Kepler's Star outshone Jupiter and reached apparent magnitude -2.5, finally disappearing in the winter of 1605. More than sheer brilliance made these two special, though their true natures were not to be recognized for another 330 years.

Another brilliant so-called nova, S Andromedae, erupted in the Andromeda Galaxy, M 31, in 1885. Others were seen in similar spiral nebulae. When astronomers finally realized in the 1920s that these fuzzy objects were actually distant external galaxies, it became clear that the erupting stars were not ordinary novae and that there was a whole new class of

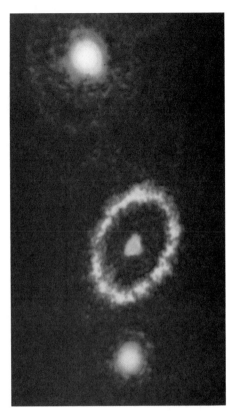

stellar violence out there, one with unimaginable power. The M 31 event was only 2.5 magnitudes—a factor of 10—fainter than the entire stellar system. These supernovae do not involve a simple thermonuclear reaction on a stellar surface, but the near or even total destruction of the star itself.

Supernovae are rare beasts. The Chinese and Japanese recorded a brilliant one in Taurus in 1054 that may have reached apparent magnitude −5. Chinese records show two other likely possibilities that occurred in 1006 and 1181. None has been seen in our Milky Way since Kepler's Star, making five in the millennium, randomly distributed. From the theory of stellar evolution, we actually expect anywhere from one to four per century in the Galaxy, but most have probably been hidden behind thick clouds of galactic dust. The local scarcity of supernovae has forced astronomers to examine them in other galaxies, but their distances are so great that we can see little detail. The only saving grace is that there are so many other galaxies that a few supernovae are discovered each year. Every astronomer alive has dreamed of seeing at least one within our own system. Thus it was with great relish that we watched a supernova blossom in the Large Magellanic Cloud in February 1987. Even though 52,000 pc away—nearby on a cosmic scale—it had such power that it reached third magnitude. Within a few days it was evident that the source was an ordinary 18-solar-mass B1 supergiant called Sk −69°202, confirming—but for a time confounding— theories of how massive stars are supposed to die.

Supernova 1987A (center) was a gleaming jewel set into the Large Magellanic Cloud near the Tarantula Nebula. The star that produced it, a B1 supergiant of 18 solar masses (left), developed an iron core that suddenly became a neutron star, collapsing with energy sufficient to blast the outer layers into space. A much-magnified view two years later (above) shows the star faded to 13th magnitude. Light from the blast had hit gas ejected earlier from the evolving B star, however, creating a great, glowing ring.

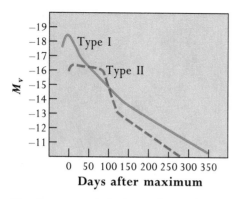

Type I supernovae (red), those with no hydrogen, are typically two magnitudes brighter than Type II (blue), which have prominent hydrogen lines and decay faster. Some Type IIs decay similarly to Type I; Supernova 1987A, though Type II, had a light curve that took nearly 100 days to reach maximum.

These wonders of the sky have been watched carefully since the 1930s, when Fritz Zwicky of Mount Wilson began to classify them, dividing them into two quite different groups. The more common, Type I, can reach an astounding absolute magnitude of −18 or −19 and decay relatively quickly after outburst. If one of them were to go off at the distance of Vega, it would equal the brightness of a hundred full Moons. Type I events also have strange spectra with no hydrogen lines, their definitive characteristic. Expansion velocities from P Cygni lines reach a remarkable 10,000 km/s, 3 percent the speed of light. Type II supernovae, a couple of magnitudes fainter, usually show a plateau in their light curves, and do have hydrogen lines; their expansion velocities are about half those observed in Type I.

The places of origin of the two types are also different. Type I events occur in galactic disks, where most of the stars reside, but also in elliptical galaxies and in the bulges and halos of spirals, showing them to be related to lower-mass stars. Type II explosions, on the other hand, confine themselves to galactic disks and to spiral arms where only the massive stars are found. So with some consternation we find Types I and II respectively associated with Populations II and I. When we look more closely, we see the types and subtypes proliferate. The most important of these is a division of Type I into the Ia, the original Zwicky class, and Ib, which although they occur in galactic arms still do not have hydrogen lines in their spectra.

Work by a generation of astronomers and physicists has allowed an understanding of how these stellar bombs work. A Type II explosion is the natural end of a massive star. As nuclear burning proceeds, the particles in the nuclei become progressively more tightly bound. Each burning stage can provide less total energy and consequently lasts a shorter period of time. It took but a million years to fuse the helium of a 20-solar-mass red supergiant into carbon. Carbon burning (which produces neon and magnesium) then takes less than 100,000 years. When the carbon is gone, the core resumes its relentless contraction and heats until the oxygen residue fires up and begins to burn into silicon and sulphur: *That takes less than 20 years.* (The process is in fact much more complex and has several intermediate steps.) Then, *in a week,* the silicon turns to iron. The temperature is over 3 billion degrees, and the reactions generate more energy in neutrinos than they do in photons. The supergiant is now layered like an onion as each stage of nuclear burning moves outward in a shell around an iron core of nearly 1.4 solar masses.

Iron is the most tightly bound of all atomic nuclei. No energy can be gained by its fusion: It is the end of the line. As the silicon-burning phase comes to an end, the core, about the size of Earth, is near the Chandrasekhar limit and is briefly supported by degenerate electrons. The iron nuclei

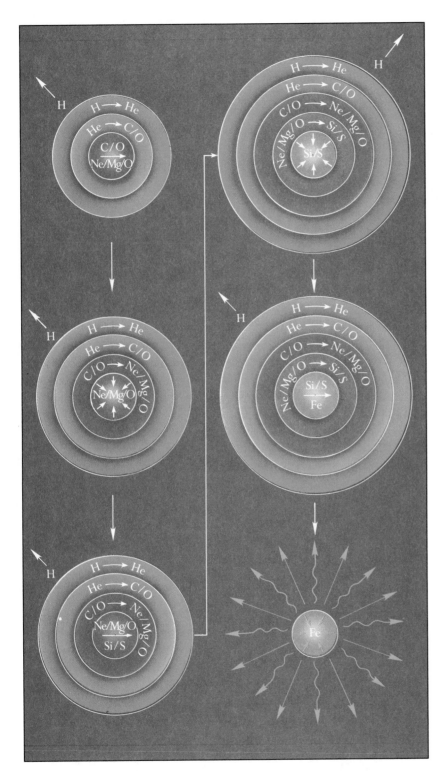

A simplified cross section of a developing supernova shows a layered structure. Carbon first burns to neon and magnesium as above. The dead Ne-Mg-O core contracts until it is hot enough to fuse into a mixture of silicon and sulfur, the silicon eventually burning to iron. The whole affair is wrapped in a huge blanket of hydrogen (the shells are not drawn to relative scale). Most of the burning takes place at the bases of the individual shells. In reality there are many intermediate isotopic products. Iron cannot fuse into anything; once formed, in a fraction of a second it undergoes a catastrophic collapse that sends a blast wave ripping through the outer stellar layers. A supernova is born.

now become subject to attack. The density is so high that the electrons start to combine with them to form manganese, and the heat is so fierce that extremely energetic gamma rays penetrate them and begin to break them back down into helium nuclei. With electron degeneracy support and gamma-ray energy both being removed from the interior, the core contracts faster and faster, then goes into a catastrophic collapse. The star has lived for 10 million years. In less than a tenth of a *second* the iron core flies inward at a quarter the speed of light to a sphere only 100 kilometers across. The gravitational energy released is beyond imagining. In that blink of an eye the star expends over 10^{46} joules—more than 99 percent of it in neutrinos. This exceeds the energy output of all the other stars in the Universe; it is 100 times more than the Sun has radiated in its lifetime.

The density near the center now becomes so great that the protons and electrons of the inner part begin to coalesce into neutrons, which condense into a ball ultimately no more than 10 to 20 km across. At the center, the temperature may initially be 200 billion degrees. The sudden implosion creates a shock wave that propagates outward. The envelope is so dense that even the neutrinos, which can normally penetrate a light-year of lead, have to fight to get out, contributing critical pressure to the blast wave that is blowing the remainder of the star apart. Only the central core is left, a shrunken star supported by the pressure of degenerate neutrons.

In the envelope's nuclear fury vast numbers of neutrons are created, which rapidly attach themselves to highly radioactive isotopes. This r-process can build heavy isotopes that the s-process of the giants cannot reach. Several solar masses of stellar material, enriched in the elements created in both the supergiant and the supernova phases, are then suddenly returned to interstellar space at several thousand kilometers per second.

Evidence for this immense nuclear activity lies in the light curve. The nuclear reactions burn the exploding envelope all the way to ^{56}Ni, a radioactive isotope with a half-life of only six days. The nickel decays into ^{56}Co, which after 57 days produces an excited nuclear state of ^{56}Fe. This isotope then relaxes into normal iron with the release of a gamma ray, which heats the gas. The star then dims in accord with the nuclear decay rates, demonstrating a huge mass of newly made metal blasting back into space.

What of the even brighter Type I supernovae? Because they appear in the old galactic halos, they cannot be caused by massive stars. Instead, we believe these events to be the final acts of white dwarfs in binary systems. There are two related possibilities. An ordinary nova is produced when a main sequence companion is tidally stretched and feeds matter onto the surface of a white dwarf, which then explodes and returns to normal. If the degenerate is close to the Chandrasekhar limit, it is possible that before a surface explosion can occur the incoming mass may push the star over the edge. Then degeneracy pressure can no longer hold it up, and the white dwarf collapses—the heat causing the whole star to ignite in a vast nuclear bomb. In a Type II supernova, the energy is generated by gravitational collapse that drives nuclear reactions; in the brighter Type I variety it is generated by the nuclear reactions themselves, evidence for which is again seen in the decay of ^{56}Co into iron. No iron or neutron core can survive; the star annihilates itself.

The other possibility involves double white dwarfs. The stars are first brought close together during red giant evolution when the expanding envelope of one embraces the other. As the white dwarfs orbit, they emit gravity waves, outward-propagating disturbances in their gravitational fields (they have been predicted by relativity theory but have never been directly observed). This emission causes a loss of energy that makes the stars slowly spiral together. Eventually, they merge as a result of tidal interaction, exceed the Chandrasekhar limit, and—as in the first model—explode. Tycho's and Kepler's stars are both thought to be Type I.

Type Ib events seem to be core-collapse supernovae like Type IIs that somehow lost their hydrogen envelopes during evolution. Possibly they were gravitationally stripped as supergiants by a white dwarf or even a main sequence companion. Such action is going on even now in the massive supergiant VV Cephei, which is passing matter onto an O dwarf. Or Type Ib supernovae may once have been Wolf–Rayet stars.

Collectively, the supernovae are partly responsible for the synthesis of the elements beyond helium, and are the sole creators of the heaviest. Supernovae may be rare, but they are potent. In the 13-billion-year lifetime of the Galaxy, maybe a quarter billion have occurred. If each releases an average of 10 solar masses (Type II more, Type I much less), the events have returned the equivalent of 2 billion Suns into space, a notable fraction of the Galaxy's total mass.

Supernova 1987A provided dramatic proof of the theories of Type II supernovae. It was first seen on February 24 by the Canadian astronomer Ian Shelton, who had been taking survey photographs of the Large Magellanic Cloud. The object was already brilliant and visible to the naked eye; it was actually photographically recorded on the rise by Robert McNaught, an Australian amateur, nearly a day earlier. But that is not the most remarkable part of the story.

Lying deep in the Kamioka zinc mine in Japan and in the Morton salt mine under Lake Erie are two massive pools of dark water designed to detect flashes of light from decaying protons. The Kamioka detector has been used to observe neutrinos from the Sun. At $7^h 35^m 35^s$ on February

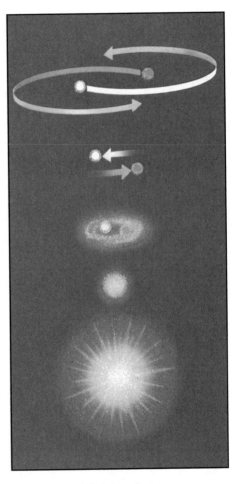

Two white dwarfs find themselves in proximity as a result of prior evolution, in which each red giant phase encompassed the other star and brought it nearer. Radiation of gravity waves (a relativistic effect) slowly draws them closer, after which a common tidal envelope drags them together. Because the combined mass exceeds the Chandrasekhar limit, the now-single star collapses and explodes, probably annihilating itself.

23, two hours before McNaught caught the supernova, the photocells that line the walls of the Kamioka detector registered that 11 neutrinos had smashed into the detector's water pool and that they had penetrated the Earth from the direction of the Large Magellanic Cloud; six seconds later, within the error of measurement for the instrument in Japan, 8 more hit the Morton mine. Given the incredibly small chance of an interaction between a neutrino and an atom, some 10 billion neutrinos must have passed through every square centimeter of the Earth (and of *you*). This number was (within the standard margins of error) that expected for a core-collapse supernova of a 20-solar-mass star 52,000 pc away. The detectors had caught the exact moment of the collapse of the core even before we saw the surface of the star respond to the event! There have been few triumphs of science to equal this one.

However, a surprise was in store. Everyone expected that the exploding star would be a red supergiant, a really massive star that had returned to the blue state, or perhaps a luminous blue variable—Eta Carinae had long been considered a prime candidate. But no one anticipated that the first nearby supernova event since the invention of the telescope would be an everyday B1 supergiant with a relatively modest mass. For a time astronomers thought that Sk −69°202 might be just a foreground star, and that a red supergiant lurked behind it. But the two-hour delay between the neutrino arrival and the beginning of the optical outburst was consistent with the relatively small radius appropriate to a B star. The light curve, moreover, was strange, not at all typical of a Type II event. Instead of coming quickly to a peak and then decaying, the star first dropped in brightness and then leisurely took nearly *two months* to come to maximum. Were the theories of internal structure and evolution wrong?

The explanation lies in the low metal content of the Magellanic Clouds. In our Galaxy this star *would* have been a swollen red supergiant, but in the Large Magellanic Cloud it had shrunk, a result of low atmospheric opacity. The small size also resulted in very high temperatures at the optical surface of the expanding debris and the production of the radiation in the invisible X-ray and ultraviolet. As the stellar surface cooled, more radiation entered the visible and the star continued to brighten to the eye in an extreme example of a decreasing bolometric correction. Four months after the explosion, a tenth of a solar mass of radioactive cobalt became visible in the debris. The light curve then dropped as the cobalt decayed into the same amount of iron. We even saw the resultant gamma rays.

The aftermath of a supernova is as extraordinary as the event itself, both in the material that is blown outward and the mass that remains behind. We look at the exploded debris first.

SUPERNOVA REMNANTS

Number one on Charles Messier's list is an easily visible fuzzy ball near Zeta Tauri, which because of its filamentary appearance is called the Crab Nebula. As early as 1921 it was identified with the great Chinese "guest star" of 1054, the brightest supernova on record, which appeared in about the same location. By 1941 the association was secure. Over the years astronomers have watched the filaments moving outward from the center at a rate of about 0.2 seconds of arc per year. To reach its current angular diameter of 3 minutes of arc, it should have begun its expansion near the year 1100. The Crab produces hosts of emission lines from which a radial expansion velocity of 1300 km/s is derived. Combination of the angular and radial expansion rates yields a distance of about 2000 pc. Allowing for some dimming of the light by interstellar dust, the absolute visual magnitude must have been around -17, about right for a Type II supernova. Over nine hundred years after the great blast, we are seeing the enriched churning debris, a supernova remnant (SNR) returning to space whence it came.

Visual remains like the Crab are scarce. To the south of ϵ Cygni is the bright Cygnus Loop, which consists principally of two arcs of filamentary gas 2° apart—the Filamentary and the Veil nebulae—and is the remnant of

The Crab Nebula, M 1, is the remnant of the supernova of 1054.

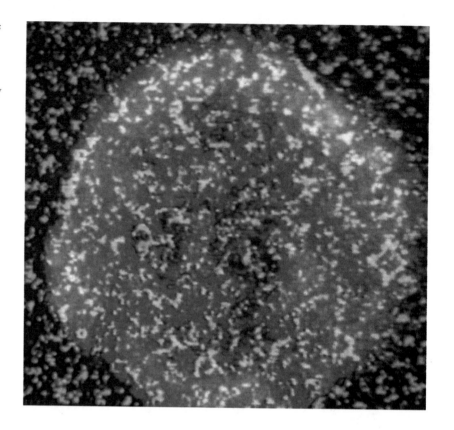

The supernova remnant from Tycho's Star radiates across the electromagnetic spectrum. Radio radiation, colored blue, is produced by the synchrotron mechanism, in which high-speed electrons are trapped within magnetic fields. X-ray radiation (green), caused by gas heated to a million degrees or more, emanates from the same region. Small wisps of optical light are indicated by red.

a star that exploded some 10^5 years ago. In the southern hemisphere, we find the huge Gum Nebula (named after its discoverer, Colin Gum) sprawled among the stars of Argo. The debris of Tycho's and Kepler's stars are far less visible, consisting of only a few expanding wisps.

Most supernovae, whether Type I or Type II, occur in the galactic plane where optical observations are seriously impeded by the absorbing effects of thick clouds of interstellar dust. To study SNRs seriously, we must use radio waves, which penetrate the dust without attenuation. In the early days of radio astronomy sources were named in order of discovery with roman letters following the constellation name; the Crab, for instance, is also called Taurus A. In contrast to their feeble optical appearances, the SNRs as a class are exceedingly luminous in the radio. Cassiopeia A, the remnant of an unobserved supernova in the seventeenth century, is one of the brightest radio sources in the sky. Some 150 of these objects are now "seen," including Tycho's, Kepler's, and the supernova of the 1006 event.

The Crab, the first clearly defined SNR, is the more uncommon of two basic kinds. It is filled with brightly radiating gas; the other kind consists of hollow shells. In the youngest ones, we see mostly the gas that was blown out of the supernova. As they age, the shock wave created by the detonation begins to sweep up ambient interstellar gas, incorporating it into the stellar material. By the time they are old, all we see—as in the Cygnus Loop—is the outline of the shock, now diminished in speed to a few hundred kilometers per second. SNR structures are, thus, controlled primarily by the prior distribution of interstellar gas. Where it is dense, the outwardly moving gas of the SNR will be retarded; where it is thin, the gas will break through. The enriched material of the explosion is then effectively merged with the interstellar medium.

SNRs were quickly found to have distinguishing radio spectra. The distribution of energy in the continuous radio spectrum of a diffuse nebula like Orion or the Rosette is relatively constant with frequency, but that from an SNR falls precipitously as frequency increases. This radiation is caused by electrons moving in a magnetic field at nearly the speed of light, reminiscent of the movements of particles in the kind of atomic accelerator called a synchrotron. As they spiral around the field lines they emit energy in the direction of their motions. Synchrotron radiation must then be polarized (that is, the waves must oscillate in the same direction), as observed. The higher the speeds and energies of the electrons, the fewer there are of them, thus explaining the drop in intensity with frequency. For most SNRs, the expanding shock wave is ultimately responsible for the radiation. It both compresses the weak ambient magnetic field of the Galaxy to a sufficient strength and provides the energy needed to accelerate the electrons. The Crab Nebula is so energetic that the synchrotron radiation extends right into the optical spectrum, where it makes a visible amorphous background to the filaments.

Optical and ultraviolet spectra generally show that the filaments radiate recombination lines of hydrogen and/or helium and also forbidden lines, mostly [O III] and [S II], typically indicating temperatures and densities of 10,000 K and 100 atoms/cm^3. In addition, we see some very highly ionized species associated with the shocks, in which temperatures extend to 1 million degrees. Even higher values are found within the hollowed-out, expanding shells. SNRs are one of the dominant lights of the X-ray sky, just as they are in the radio domain, a remarkable testimony to their huge energies. The X-ray emission shows heated gas with temperatures over *10 million degrees*. The Crab, however, is different. Its X-ray spectrum is dominated by the synchrotron mechanism, showing electrons moving at near 99 percent the speed of light.

This incredibly hot gas is expanding away into interstellar space. Long after the SNR has dissipated, a huge, hot bubble remains. Eventually it connects with expanding bubbles from other SNRs, and they weave a tapestry of hot loops and tunnels, its temperature measured in the hundreds of thousands of degrees, that pervades the entire Galaxy. The gas has the power to compress the interstellar medium and thus seems to be one of the triggers that can begin the formation of stars, the old creating the new.

PULSARS AND NEUTRON STARS

In Type II events, the iron core collapses inward, a phenomenon predicted as long ago as 1934 by Baade and Zwicky. Baade even tentatively identified the stellar remains in the Crab with a 13th-magnitude star that has no absorption lines, but he had no way of proving that the two were actually connected; that had to wait until 1967.

Anthony Hewish, a British radio astronomer, had designed a telescope to look for rapid changes in the signal strengths of pointlike radio sources, "twinkling" caused by the passage of radio waves through the solar wind. A graduate student working on the project, Jocelyn Bell, found instead a strange source in the constellation Vulpecula that turned on with a sharp pulse that recurred every 1.337011 . . . seconds. Sometimes the pulses faded away and were completely invisible for months. She noted, however, that when they returned, they occurred at just the same intervals as before. That the source was celestial in origin was quickly obvious from its passage across the sky in a sidereal day. Even a distance could be measured. Light and radio waves travel at c only in a vacuum. In a medium they go more slowly, which is the cause of refraction. Thin as it is, the ionized interstellar gas slows radio waves and also disperses them, lower frequencies having slower speeds. Sure enough, the lower the frequency, the later the pulses were seen to arrive. An estimate of the density of interstellar space put the source—designated a pulsar—300 pc away.

Bell and Hewish knew *where* the pulsar was, but not *what* it was. Was the mechanism producing such regular signals natural, or was it evidence of another distant civilization, perhaps some kind of beacon? The only physical system that could account for such regularity was rotation, but what could be rotating fast enough to produce the observed period? The answer lay in the prescient prediction of neutron stars by the Russian physicist Lev Landau in 1932 and by Baade and Zwicky in 1934. To rotate that fast, a body would have to be very small, only 10 or so kilometers in diameter— just the size expected for the collapsed stellar remnant of a supernova. The pulsar must be a degenerate neutron star that had well over a solar mass

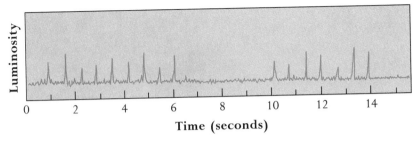

A typical pulsar radiates short, sharp, regular pulses with no radiation between (except an occasional interpulse, not shown here). Sometimes it turns off completely, but when it returns the pulses are right on schedule.

condensed within the size of a small town spinning somewhat faster than once per second.

The clincher came when a pulsar was discovered in the Crab Nebula, one with a record-setting short period of 0.03106 seconds, clearly relating the pulsars to supernovae. The search was finally narrowed in the optical when astronomers examined the candidate stars in the Crab's center with a rapidly recording photoelectric photometer. One star was seen to glow brightly for a fraction of a second, then *disappear*. This was the one predicted by Baade 50 years earlier.

The scientific journals quickly filled with more discoveries about these odd characters, as their number grew to the 400 or so known today. The periods have a wide distribution, from that of the Crab up to about 4 seconds. Furthermore, they are not entirely regular; individual pulses can fall slightly off the average and flicker about the mean. Most pulsars are slowing down, albeit at a very slow rate. The Crab increases its period by about a ten-thousandth of a percent a day. There are also glitches, wherein the pulsar will suddenly speed up a bit only to resume its steady decay. The masses (derived from orbital effects in binary pulsars) are around 1.4 $M_\odot$, close to the value expected if their progenitors are supernovae.

We now have a good general explanation for all these phenomena. A normal atom is mostly empty space. Only one part in 10^{15} is actually filled with protons or neutrons. A neutron star of 1.4 $M_\odot$ only 10 to 20 km across can exist because the space is nearly squeezed out, resulting in a density near 10^{12} gm/cm^3, a million times denser than a white dwarf. You would attain the same density by shoving the mass of the Earth into a typical sports stadium. The rapid spin of the pulsar owes itself simply to the conservation of angular momentum. As the radius of the collapsing star decreases after a supernova, its spin velocity increases. Moreover, the mag-

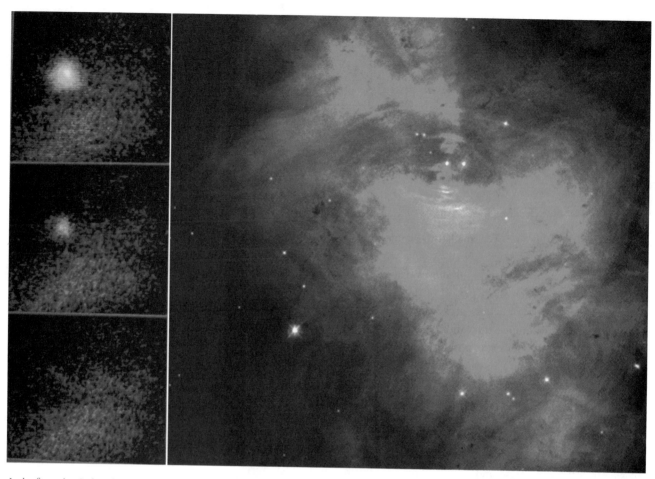

Light from the Crab pulsar, visible over the entire electromagnetic spectrum, is sampled at left in X-radiation at 3 different moments in the cycle. At right, Hubble shows us the pulsar's powerful effect on the surrounding gas, a scene that changes over a period of months.

netic field of the one-time star is collapsed along with the matter, attaining a strength some 10^{13} times that of the Earth.

The magnetic field axis of a pulsar, like the Earth's, is oblique to the rotation axis, so that it wobbles around. The moving magnetic field is thought to produce a powerful electric field that accelerates electrons along the magnetic axis at a speed nearly that of light. The result is a tightly concentrated beam of electromagnetic radiation that revolves as the star spins, appearing something like a skewed lighthouse beacon. If the Earth happens to be in the way, we receive a pulse of radiation; if not, no star is perceived at all. If the angle between the magnetic axis and rotation axis is high, we may catch a glimpse of the other pole as well as it goes past, providing us with an interpulse, such as we receive from the Crab.

The radiation derives its energy from the stellar rotation. Consequently, the pulsar must slow down with time; the youngest, like the Crab, rotate the fastest and have the highest energies, explaining why the Crab pulsar is seen not only at optical wavelengths but in the X-ray as well. Indeed, the pulsar is thought to be the source of the magnetic field and high-speed electrons that give the Crab its extraordinary power. Only the youngest pulsars will be associated with SNRs because the expanding clouds dissipate long before the spinning star runs out of energy. Most SNRs have no visible pulsars, either because the star was entirely destroyed (in a Type I blast, for example) or because the rotation axis is not oriented to bring the beam past the Earth. It is also possible that some pulsars are produced in a stellar collapse that does *not* generate a supernova. When all the possibilities are considered, the number of observed pulsars is consistent with the galactic supernova production rate. Together, the supernovae and their stellar and gaseous remnants may be the origin of the mysterious cosmic rays, high-speed atomic nuclei constantly crashing into the Earth with energies that dwarf those created in our greatest atomic accelerators.

The observed glitches are a result of the strange internal construction of a pulsar. The little star is no longer gaseous. Its surface is thought to be a crystalline solid. Inside, the matter is in a superfluid state: that is, it has no

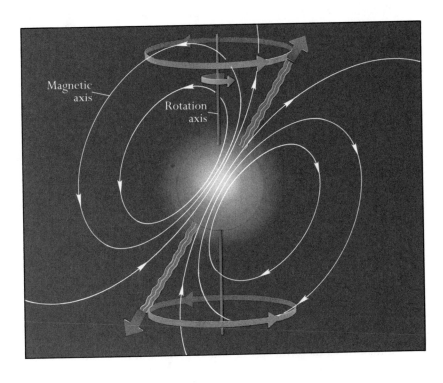

A pulsar's magnetic field axis spins wildly about, generating electromagnetic radiation directed along the axis; if the Earth is in the way, we receive a sharp blast of energy.

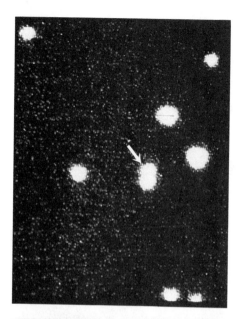

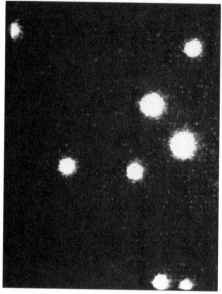

Top: *The Black Widow pulsar's companion heated on one side by fierce blasts of radio radiation, and is visible (see arrow). Bottom: When it eclipses the pulsar, the radio beeps vanish. The companion's heated side is turned away from us and it disappears too.*

viscosity, no friction within itself. The interior can be spinning at a rate different from that of the crust. Vortices within then interact chaotically with the crust to force the outside temporarily to spin faster. Still, our ignorance runs deep; the various theories are only partial ones, and no one yet knows the actual mechanism for the production of the beamed radiation from the magnetic field.

If this picture is not strange enough, consider another breed, the millisecond pulsar. The fastest known rotates an astonishing *885* times per second, about 30 times faster than the Crab. The explanation—as so often for bizarre behavior seen in normal stars—is stellar duplicity. The more massive member of a binary system explodes in a supernova and becomes a pulsar, which slowly decays away. The companion survives and eventually begins its own evolution. As it expands, it sends matter raining down onto the old pulsar, providing a kick that makes the pulsar spin faster, to extraordinary speeds. The energies are now so high that we see a rare millisecond X-ray pulsar, which becomes a millisecond radio pulsar after the companion has evolved into a white dwarf. We can see the effect of such a companion as orbital motion produces a varying Doppler shift in the pulse arrival times.

There are, however, two millisecond pulsars clearly *without* binary companions, apparently negating the theory. A third pulsar solves the mystery. This one is an *eclipsing* system, and from the observations of its orbit we can see that its companion has both a very low mass and a large radius. Apparently, this pulsar—dubbed the Black Widow—is in the process of consuming its mate, as the two lone millisecond pulsars have done already.

BLACK HOLES: THE ULTIMATE ENDS

If you throw a ball upward, it loses kinetic energy under the action of gravity, slows, and falls back to Earth. If you could throw it upward at 11.2 km/s, the terrestrial escape velocity, the strength of the Earth's gravitational field is insufficient ever to bring it to a stop. Although the ball would perpetually slow, it would go on forever into space, never returning.

In 1916 Albert Einstein showed, in his general theory of relativity, that mass—the source of gravity—curves spacetime. One of the fundamental precepts of relativity is that the speed of light in a vacuum is a constant, independent of the speed of the source or observer. Since spacetime is curved, a ray of light, although massless, must be affected by gravity in a way similar to the ball. Shine a flashlight upward; the beam must lose energy, but its speed, unlike the ball's, will stay constant. Since the energy of a photon is $E = h\nu$, the energy will instead be lost via a decrease in frequency. The wavelength gets longer, and the light reddens.

Now, in your mind, move into space and look back at the Earth, from which a companion shines a green light directly at you. Because the Earth's gravitational field is weak, the light will be reddened only imperceptibly. At the same time, however, your friend begins to compress the Earth. The gravitational field at the surface goes up, as do the escape velocity and the degree of reddening. By the time the Earth is compressed to a little less than 2 centimeters across, the escape velocity reaches c: Light loses all its energy, the wavelengths are shifted to infinity, and both the flashlight beam and the Earth itself disappear from view. Our planet has fallen into a black hole, a puncture in spacetime, from which nothing can ever escape. Anything in that condition must collapse forever, the mass eventually reaching infinite density at the center. The laws of physics within the black hole have lost their meaning, because no information can ever exit. However, as the black hole collapses, it leaves behind an apparent surface called the event horizon, the limiting radius at which the strength of gravity just barely stops the light from escaping. The size of the event horizon, which is taken as the size of the black hole, depends only upon the black hole's mass and not upon its degree of compression, so it stays constant over time.

Although the theory of black holes was first worked out in 1916 by Karl Schwarzschild, it was not until the last 20 years or so that astronomers began to find objects in the sky whose explanations seemed to demand it. The phenomenon seems so bizarre as to be impossible; but the conditions

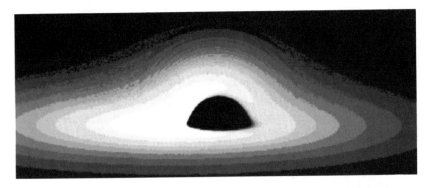

A black hole can be likened to a puncture in spacetime. No radiation, no information, can escape from within the event horizon. The hole may be bright around its periphery if matter is flowing into it, but the interior is a mysterious place where the laws of physics no longer have any meaning.

are not that far from those of neutron stars, which we *know* exist. Support by degenerate neutrons has a limit just like the Chandrasekhar limit of a white dwarf. Above about 3 solar masses, a neutron star can no longer support itself and must collapse to a black hole. The event horizon is a few kilometers wide, not much smaller than the neutron star.

But do black holes really exist? If they do, where are they? And if no radiation can escape, how can we ever hope to find them? The key lies in the same phenomenon that produces novae, symbiotic stars, and (possibly) Type I supernovae: tidal interactions. In the constellation Cygnus is an otherwise ordinary O (or possibly B) dwarf that radiates powerfully in the X-ray spectrum. The star, called Cygnus X-1, displays absorption lines that Doppler-shift back and forth, showing it to be a member of a binary system—yet there is no trace of its companion.

If we know the velocities of both members of a spectroscopic binary as well as the inclination of the orbit to the line of sight (as observed from an eclipse), then we can determine the stellar masses from Kepler's third law. Velocities from a double-line spectroscopic binary without the inclination yield the mass ratio and lower limits on the individual masses. A single-line version gives even less information, only a complex function of masses and the inclination. But if we can estimate the mass of the O star in Cyg X-1, then we can at least place a lower limit on the mass of the invisible companion. It turns out to be *over 3 solar masses,* and it may be close to *16.* Any normal star of that mass should be visible. It is too massive for a white dwarf or a neutron star, and therefore we expect it to be a black hole. The X-rays are caused by material that overflows the tidal surface of the O dwarf and enters an accretion disk that is heated to a high temperature before disappearing forever down the celestial drain. Without the X-rays we never would have noticed the star in the first place. There are two similar candidates: LMC X-3 in the Large Magellanic Cloud and the obscure A0620-00 in Monoceros, both of which seem to be closer to 10 $M_\odot$.

But in spite of their enormous X-ray luminosities, these three black hole candidates are relatively normal compared with one called SS 433, which has one of the strangest spectra known anywhere. Superimposed on an ordinary set of absorption lines are twin sets of Doppler-shifted hydrogen and helium emissions that wander across the spectrum, criss-crossing back and forth with a period of 164 days. They imply material that is alternately approaching and receding with a speed *at least one-sixth that of light.* From the small Doppler shifts in the absorptions, we see a B star in orbit about an invisible companion. Apparently, it is again sending mass onto a collapsed star of some kind.

Cygnus X-1 appears to reside quietly among its neighbors in the Milky Way. It is one of the most powerful X-ray sources in the sky, and its radiation demonstrates that matter is flowing into a collapsed object—an invisible companion of such great mass that the O star must be orbiting a black hole.

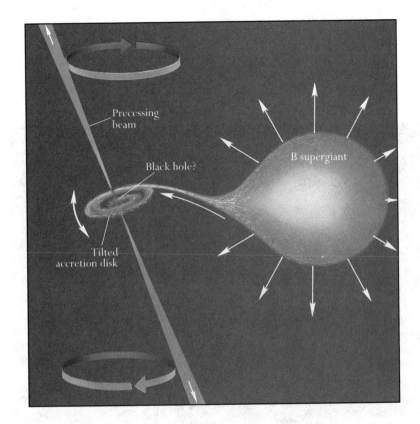

Precessing
beam

Black hole?

B supergiant

Tilted
accretion disk

The collapsed object—a black hole or neutron star—in SS 433 accretes matter from a class B companion, then sprays it out from its poles. The accretion disk precesses with a 164-day period, causing the twin beams to oscillate wildly.

Not all matter that flows into SS 433's accretion disk falls into the central gravitational source. Some of it leaves in violent twin jets from the rotation poles. Subtle Doppler motions show that the accretion plane, and the poles, are precessing at the 164-day period, the jets spraying wildly around like a garden hose gone berserk. Radio telescopes even show us the corkscrews of matter spurting away. The mass of the source of this extraordinary behavior is close to the black hole limit; moreover, it is inside a known supernova remnant called W 50, which is lit in the X-ray by the ferocity within. SS 433 may well be a neutron star. Nevertheless the evidence for the existence of black holes is quite strong. In addition, at the centers of galaxies, we see huge masses packed into probable black holes, sometimes with jets of matter shooting out in opposite directions, a phenomenon to be explored in the next chapter.

So far we have seen an escalation of stellar strangeness, from the apparently simple Sun, through the evolution of the giants and the supergiants, to the end products—the dense collapsed white dwarfs, neutron stars, and black holes. Now we come to perhaps the most remarkable part of the story: the beginning.

Barnard 86—a dark cloud and stellar birthplace—is silhouetted against the Milky Way.

Chapter Seven

FIRST LIGHT OF DAY

THE BIRTH OF STARS AND OF OURSELVES

It all began, apparently, with a big bang. Thirteen billion years ago, maybe 10, maybe 20, all the matter of the Universe was concentrated nearly (perhaps *really*) into a point. We start counting at 10^{-43} seconds after the event itself, the smallest scale of time allowed by Planck's constant. At this moment all the forces of nature were unified, but they quickly began to decouple as the rapid expansion lowered the density. At 10^{-33} seconds, matter as we know it appeared, and by the time the Universe was a second old, the density had dropped enough so that the matter finally (!) became transparent to neutrinos.

Our stellar story really begins when the Universe is about three minutes old. Its temperature and density are rather like the interior of a star, allowing some of the newborn protons to begin to fuse into deuterium and then helium, building the latter to about 8 percent (by number) of the count of hydrogen. After this fury of activity, 100,000 years passes until the next stage, when the heat is low enough to allow electrons to combine with nuclei in the construction of complete atoms. Photons of radiation, no longer thickly bound to matter, begin to roam free. As the Universe cools during the next billion or so years, instabilities in the expanding gas grow to produce gravitationally bound blobs of matter, nascent clusters of galaxies. These fragment into smaller lumps that mature into the galaxies, and these in turn eventually precipitate the stars, the planets, and finally ourselves.

THE STATE OF THE UNIVERSE

Today we look into space and see the results of the grand event that took place so long ago. We live within a sparse cluster of about 30 galaxies parochially called the local group. It is anchored at one end by the Milky

The most basic unit of mass in the Universe is probably the cluster of galaxies. This one, in Hercules, has been gravitationally bound together since its formation some 13 billion years ago.

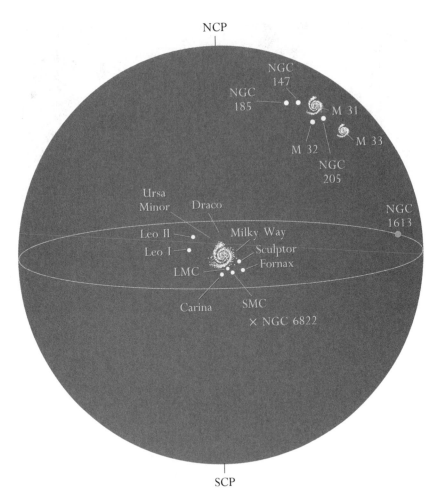

NCP

NGC
147

NGC
185

M 31

M 33

M 32

NGC
205

Ursa
Minor

Draco

NGC
1613

Leo II

Milky Way

Leo I

Sculptor

Fornax

LMC

Carina

SMC

× NGC 6822

SCP

Several members of our local group of galaxies are depicted in a three-dimensional projection; the viewer is looking roughly in the direction of 6ʰ right ascension. The small galaxies tend to clump about the two major components, the Milky Way and M 31.

A close-up view of the giant elliptical galaxy M 87.

Way system and at the other by our near twin, the Andromeda spiral, M 31, about 750,000 parsecs away. At roughly the same distance is the lesser Triangulum spiral, M 33. The field is rounded out by numerous small assemblies. We are accompanied by the Large and Small Magellanic Clouds, classified as irregular galaxies, those with little form or shape and masses only about 1 percent of our Milky Way. Similarly, M 31 is escorted through space by the small elliptical galaxies M 32 and NGC 205. Elliptical systems

The Leo I dwarf elliptical is so puny it barely seems to merit the term galaxy.

Stephan's Quintet is a small, compact group in which the galaxies are so close they interact tidally. The spiral to the upper left may not be a member of the system.

have no disks or spiral arms. They are nearly pure Population II, devoid of significant interstellar matter and young luminous stars. Surrounding the three major galaxies are several low-mass dwarf ellipticals, some of which are so faint they can barely be detected.

Scattered into the distance as far as we can see are vast numbers of galactic groupings. Some, like our local group and Stephan's Quintet, are sparsely populated, but others—like the Hercules cluster—are hundreds of times richer. As we gaze farther afield, other varieties of galaxies are to be seen. A normal spiral—like our own, M 31, and M 33—has its arms emanating directly from the central bulge. But here and there we also see barred spirals, in which the arms emerge from a bar of stars that pierces the nucleus like a knitting needle punched through an orange. Rich clusters of galaxies frequently possess giant ellipticals, galaxies vastly larger than M 31's poor neighbors. In the nearest substantial cluster, the great Virgo Cloud 16 megaparsecs away, the spectacular M 87 system possesses more than 10^{13} solar masses, over 10 times the mass of the Milky Way. Occasionally we also see starburst galaxies, irregular masses rich in interstellar gas and dust in which star formation proceeds at an extreme rate.

It is not difficult, as seen in the photographs in this and earlier chapters, to resolve the nearer galaxies into familiar objects: Cepheids, O and B stars, planetary nebulae, novae, and so on. Since their absolute magnitudes are known from studies of our own system and the Magellanic Clouds, we need only measure their apparent magnitudes to derive distances. When the galaxies are too far away for us to see individual stars, we might then use their globular clusters or occasional supernovae. And when nothing can be resolved, we can at least make the measure on the basis of galaxy type and total apparent brightness, where the absolute brightness is calibrated from the closer objects.

The overriding feature of the Universe of galaxies, however, is neither content nor distance, but *motion*. As long ago as 1917, before anyone knew what galaxies were, V. M. Slipher of the Lowell Observatory in Arizona saw that most had their spectral lines Doppler-shifted to the red, indicating that they are receding at remarkable speeds. After Edwin Hubble made the first measurements of distances to the galaxies, in the 1920s, and proved that they were outside the Milky Way, he realized that their separations from us are strictly coupled to their recession velocities. The more distant the galaxy, the faster it is moving. Moreover, the relation is linear: If the distance is doubled, so is the speed. This relation, just what would be anticipated as the result of an explosion, is the principal proof that the Universe began in a Big Bang and is ever expanding. Further dramatic evidence lies in the discovery that we are bathed in radio radiation from a blackbody only 3 degrees above absolute zero, just the level expected from a primeval fireball that has had 10 to 20 billion years to expand and cool.

However, while the explosion analogy is useful, it is deceptive. The galaxies are not expanding into space like the fragments of a dynamite blast, but define space itself. It is the Universe, which has neither edge nor center, that is expanding. It is getting larger and is carrying the galaxies along with it, the distances between them increasing much the way raisins separate in a rising loaf of holiday bread. If you could transport yourself to a distant system, the view would be the same, the galaxies moving farther and farther away. Strictly speaking, the relation applies not to individual galaxies, but to *clusters* of galaxies. The galaxies within a group are bound by the force of their own gravity. The local group is getting no larger; it is simply separating from all the other clusters.

The primary number of the Universe is the Hubble constant (H_0), the expansion rate measured in kilometers per second per megaparsec. The Hubble constant leads directly to the age of the Universe, most simply defined as the time it took any one galaxy to get from us to its present position at its measured velocity. The age must be the same for all galaxies because the velocity–distance relation is linear and is the reciprocal of H_0, called t_0, the Hubble time. This value is an upper limit to the actual age,

The barred spiral NGC 1365.

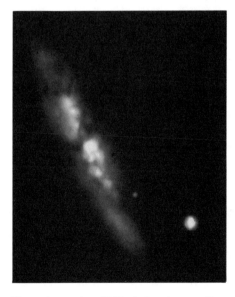

The starburst galaxy M 82, thick with interstellar dust; individual stars can barely be seen.

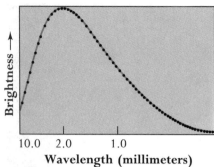

The background radiation of the Universe is, as far as we can tell, a perfect blackbody at 2.735 K.

The distribution of the temperature of the background radiation across the entire sky: Red indicates cooler than average, blue hotter. The dipole appearance is the result of the motion of our own galaxy relative to the smooth expansion, caused by local gravitational effects; when that is removed, the distribution of the background radiation is perfectly isotropic (the same in all directions).

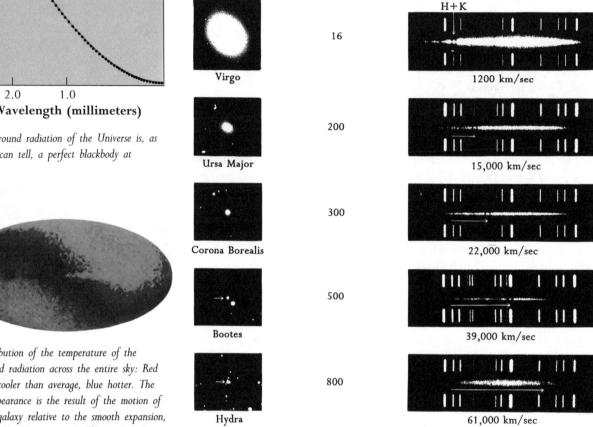

Member of cluster in	Approximate distance in megaparsecs	Red shifts and velocities
Virgo	16	1200 km/sec
Ursa Major	200	15,000 km/sec
Corona Borealis	300	22,000 km/sec
Bootes	500	39,000 km/sec
Hydra	800	61,000 km/sec

Photographs and spectra of galaxies show that distance is closely allied with red shift and thus with recession velocity. The rate of expansion is taken as 75 km/sec/mpc.

because gravity must slow the expansion, but the Hubble time is expected to be within 30 percent or so of reality.

It would seem to be a simple affair to determine H_0 and t_0 from the distance and recession speed of any one galaxy, but it turns out to be quite difficult. Even the closest systems outside the local group are hard to resolve; they are also subject to gravitational influences from other galaxies and groups, which produce significant local deviations from the smooth Hubble relation. If we look much farther away, where the velocities are greater and local motions not so significant, distance methods are no longer so reliable. Using various techniques, astronomers find H_0 equal to anywhere from 40 to 100 km/sec/mpc, rates that respectively give Hubble times of 25 and 10 billion years. We seem slowly to be converging on a value in the middle, about 65 km/sec/mpc, which makes t_0 equal to 15 billion years, and the actual age somewhat less. The slowing effect of gravity, which makes the Universe younger than t_0, has made the globular clusters seem older than the Universe, a problem solved by an *expansive* force, first proposed by Einstein, that counters the gravitational pull. New and better parallaxes, and the distance methods calibrated by them, as well as new calculations of stellar evolution, appear to be reconciling the two ages. But there are no certainties in this subject. The arguments still rage and are likely to do so for some time.

Sprinkled among the normal galaxies are a variety of oddballs. For example, the giant elliptical galaxy M 87 displays a straight jet of matter that extends outward from its core for nearly 2000 parsecs. The Universe displays quite a range of these active galaxies. A favorite is Cygnus A. An optical photograph shows only a faint peculiar elliptical galaxy; however, in the radio domain it is at the center of opposing narrow twin jets that extend nearly *half a million* parsecs into space, where they blossom into huge clouds of low-density gas. We are reasonably certain that the central engine of this awesome power is a black hole, one that might reach a billion solar masses. Gravitational interactions among the stars in the center of such a galaxy may cause individuals to spiral into a black hole, where they are ripped apart, some of the matter spraying clear out of the galaxy in an event reminiscent of SS 433 (Chapter 6), but on a gigantic scale.

The quasars, an acronym for quasi-stellar radio sources, are the pinnacles of these active systems. Many of the early radio sources had no apparent optical counterparts. In the early 1960s, astronomers discovered that several were associated with what at first appeared to be only ordinary stars. But the "stars" had emission lines at strange wavelengths that Caltech's Maarten Schmidt found to be hydrogen Doppler-shifted by enormous amounts. If we assume that a quasar partakes of the expansion of the Universe, then its distance can be estimated from velocity and the Hubble constant; the result implies that they are the farthest things known in the Universe, up to several billion parsecs (and light-years) away. From their apparent magnitudes, we then find that quasars must be some of the *brightest* objects in the Universe as well, a thousand times more luminous than the Milky Way, all the brilliance in each compacted into a tiny point in the sky. Some also have associated jets, and most are actually radio-quiet. Our only explanation is, again, the creation of energy by the infall of matter into

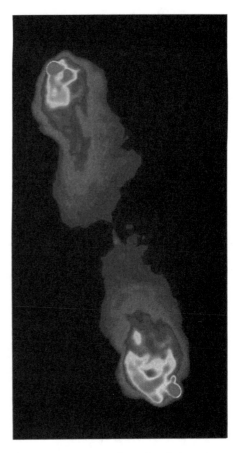

Cygnus A—a tiny, peculiar galaxy between the two lobes—unleashes awful violence into deep space via huge, tightly collimated twin beams.

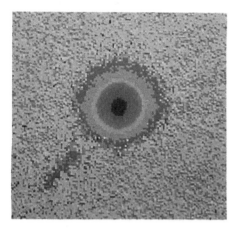

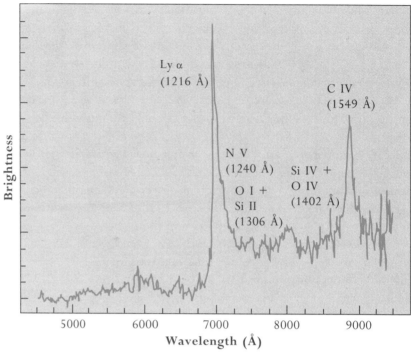

The quasar 3C 273 (above) displays a narrow jet; its spectrum shows that the quasar is moving away at 16 percent c—a small increment compared with that of PC 1158+4635, whose spectrum is at right. One of the most distant quasars, it is moving so fast that the ultraviolet spectrum is shifted into the red. At these velocities, the Doppler shift is no longer linear with speed.

a massive black hole. There are astonishing numbers of quasars, hundreds per square degree of sky. Their distances are such that in perceiving them we are looking significantly back in time. If the quasars are several billion light-years away, we see them as they were several billion years ago. They may represent the earliest stages of galaxy formation.

DARK PROBLEMS

So far it sounds as if all is reasonably well understood; but there are severe problems. Some astronomers claim that the quasars may not be at the great distances implied by their speeds and the Hubble relation. In some cases galaxies appear to be connected to other systems—including quasars—that have very different red shifts, suggesting that another mechanism is operating to displace the lines toward the red. The large spiral to the upper left in Stephan's Quintet has a red shift different from that of the other four members. Is it really part of the group? More disconcertingly, the Universe is exceedingly lumpy, the uneven distribution of its galaxies quite inconsistent with the perfectly smooth distribution of the 3° background radiation.

Worse is the problem of mass and deceleration. Gravity must drag the Universe back and slow its expansion. If there is not enough matter, the Universe will expand forever and can be considered open. However, if the average, smoothed-out density is greater than a critical value of about 10^{-29} gm/cm^3 (the exact value depends on H_0), the Universe is closed; the expansion will eventually stop, to be followed by contraction and a "great crunch." If the density is exactly the critical value, the Universe is still closed and the expansion will coast to a halt, but only after an infinite time. Because we look back into time when we look out into space, we ought to be able to measure the deceleration rate: The more distant galaxies should be moving away faster, since they have had less time to decelerate. But that measurement is only now becoming possible.

A more practical approach would be simply to measure the mass of a significant sample of the Universe and divide by the volume to find the mean density. The most straightforward way is to count stars, or rather to estimate mass by the total luminosity of the galaxies within a certain volume, making a correction for the amount of known interstellar matter. The

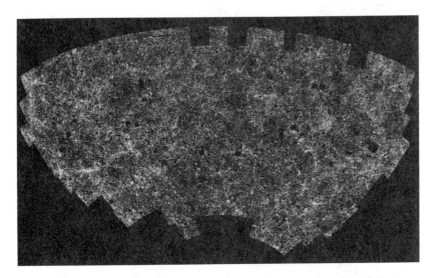

The positions of 2 million galaxies are plotted around the south galactic pole. Individuals are grouped into clusters, clusters into superclusters and even larger structures, all with great voids in between.

ratio of the critical mass of the Universe to that observed is called omega, Ω, and this method yields a value of about 0.01, far short of the amount needed for closure. Thus the problem is apparently solved, and we live in a one-way Universe that goes on forever.

But does it? It is also possible, indeed preferable, to derive mass from its gravitational effects. The individual members of a cluster of galaxies all mill about one another. The galaxies are too far away for us to witness actual orbital motion, but the radial velocities of the cluster members relative to the average must still be responsive to total mass. The gravitationally implied mass is some 10 times that indicated by the galaxies' collective brightness. Something is there, pulling on the galaxies, but we cannot see what it is.

The same effect is visible even within a galaxy. Its stars are all in a state of rotation about the center. Orbits in the disk are fairly circular, those in the halo more elliptical. Orbital periods must be controlled by Kepler's third law as generalized by Newton, that is, the period at any given distance from the center must depend only on the amount of mass interior to the orbit. Find a spiral and measure the radial velocities along the disk relative to the center. The result, after correction for the tilt of the plane, is the galaxy's rotation curve. Application of Kepler's third law gives the amount of mass interior to all the points along it.

What we find is remarkable. Move outward along the galaxy's radius. Even after the luminosity has diminished nearly to zero, *the mass keeps increasing.* Radio astronomers come to a similar conclusion by observing emission lines radiated by interstellar matter. Exactly the same effect is seen in our own Milky Way, where, as we will see, the rotation curve is established by a combination of radio and optical studies. There are 150 billion solar masses interior to the solar orbit 8.5 kiloparsecs out from the center. At

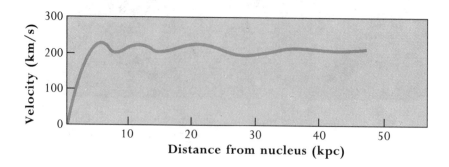

The rotation curve of NGC 801. Dark matter is making the stars move faster than anticipated from the number of stars.

twice that distance there is twice the mass even though the amount of visible material steadily declines. So much mass is out there, we ought to be able to see it. But we cannot. The mysterious invisible stuff is referred to only as dark matter (not to be confused with dark *known* interstellar material).

A powerful astrophysical argument yields about the same result. The ^{2}H (deuterium), ^{3}He, and ^{7}Li in the Universe (discounting what are in stellar cores, which are inaccessible) were all made in the early moments of the Big Bang. The abundances of these isotopes depend on the density at the time of creation, which is consistent with the present density derived by gravitation methods. The final result of all the studies is that there seems to be 10 times as much matter in the Universe as determined on the basis of simple luminosity. Omega is then raised to 0.1. Although we can see only a tenth of what is there, the Universe still seems quite open.

Enter the theory of the Big Bang. In order for Ω to be 0.1 now, it had to be within 10^{60} of *exactly* 1 at the start of our count of time. Any greater deviation would have led to current values of Ω different from that observed by many factors of 10. Thus there are theoretical grounds for believing that Ω must be *exactly 1;* that is, that the Universe is closed, but just barely and on an infinite time scale. If that is the case, there is a *hundred* times more mass in the Universe than we can see. Only a tenth of *that,* moreover, is actually in the galaxies and their clusters, producing gravitational effects within and among them. The rest is somehow floating freely throughout the Universe—and not in any ordinary atomic form, since it did not participate in the initial fusion of hydrogen into ^{2}H and ^{3}He.

We do not know the forms of the dark matter. To solve a part of the problem, huge numbers of brown dwarfs have been invoked, but to date, few have responded to the call; there are nowhere near enough of them. Perhaps the Universe is filled with black holes left over from the time of creation. Exotic subatomic particles may yet remain to be found. Maybe neutrinos have a tiny amount of mass; there were so many created in the Big Bang that even if each has only an infinitesimal amount, their sum could close the Universe and satisfy theory. But if they have mass they cannot travel exactly at the speed of light, and the timing of the neutrinos from supernova 1987A relative to its visual sighting suggests that neutrino masses are insufficient.

The existence of cold dark matter does provide some succor, as its gravitational action may make the lumpy distribution of visible matter consistent with the smooth background radiation. That is small relief. The fact is that we are trying to create theories about the Universe on the basis of the small part of it that we can see.

BIRTH AND YOUTH OF THE GALAXY

Within this uncertain context we now embark on a seemingly simpler quest, seeking knowledge of the formation of our own Galaxy and of the stars within it. Imagine the murky galactic beginnings, at least as we have been able to piece them together. A spherical rotating mass of hydrogen and helium gas approaching 10^{12} $M_\odot$ separates from the great initial cloud that has begun to give birth to our local group. Ever so slowly, the huge assembly contracts under the force of its own gravity, the density finally increasing to the point where turbulent eddies can stay together, developing the first generation of stars.

Since Big Bang theory shows clearly that only helium, deuterium, and lithium could be made by fusion in the first few moments of the Universe, these newborn stars could have no metals. But the oldest stars that we know, those in some of the globular clusters and a few in the halo, all have *some* metal content, even if only 10^{-4} that of the Sun. No truly metal-free star has ever been seen. There is therefore a gap between the earliest observational record and the birth of the Galaxy that we have not yet filled. Where is the first generation, sometimes dubbed Population III? It may well be out there. Stars very poor in metals are quite scarce, and the members of Population III would be rarer still. The high-mass luminous ones long ago burned themselves out, so all that would remain would be the lower main sequence from cool class G on down, which would be hard to discover. Only a few Population III supernovae would be required to sow the Galaxy with enough metals to produce the abundances seen in the most metal-poor stars, proportions satisfyingly consistent with the explosive r-process. Alternatively, early conditions may have led to the exclusion of low-mass stars altogether. The absence of this earliest generation does not then necessarily pose any difficulty. Nevertheless, we would feel more secure if we could locate just *one*.

The first generation is gone. The Galaxy is now contracting and fragmenting into huge clumps of gas that will become the oldest globular clusters. These in turn splinter into stars. A young globular with its complement of O and B stars, now long departed, must have been an extraordinary sight. Tidal forces induced by the mass of the rest of the Galaxy probably disrupted many of the clusters and helped seed the halo with the individuals we now see in the general field.

Since many globular clusters were formed while the rotating, birthing cloud was contracting and in a state of chaotic motion, they had a component of velocity toward the galactic center, resulting in the elongated, elliptical orbits they retain to this day. The current spatial distribution of these great assemblies retains something of the memory of the ancient times when the Galaxy was more or less spherical. Many of the clusters are easy to see, as they lie outside the thick, obscuring dust of the galactic disk. Their distances are also fairly simple to measure from the brightnesses of their stars, allowing the determination of the centroid of their distribution. Thus it was that Harlow Shapley could find the distance to the center of the Galaxy some 70 years ago. The currently accepted value from all halo sources is 8.5 kpc.

The subsequent development of the Galaxy is the subject of considerable controversy and research. The basic idea is that the star-forming interstellar matter continued to contract while the new stars born within constantly enriched the gas in metals as a result of stellar evolution and mass loss through planetary nebulae and supernovae. As a result of conservation of angular momentum, the Galaxy's rotation speed increased, spinning it out into a disk that continued to thin to the one we see today. Each new generation of stars, then, had more circular orbits about the galactic center and was made from raw material that had more heavy elements. We therefore see something of a relation between galactic orbits and stellar abundances, explaining why Population I, which includes the Sun, is more metal-rich than Population II. The relation between galactic kinematics and abundances first described in Chapter 3 is thus nicely explained.

However, reality could not have been anywhere so neat as this simple picture. The globular clusters are hardly a monolithic class. Although all are surely old, they are not all the same age. The positions of their main sequence turnoffs indicate a range of perhaps 6 billion years. Metallicities also exhibit a large range, from a low of a hundredth or so that of the Sun to maybe a third.

Moreover, there seem to be two populations of globulars, one that spreads out into the halo, and another, the higher-metal variety, that is confined to a thick disk. The halo seems to have taken much more time to collapse than the simple picture would suggest, and the relationships between metallicity, age, and galactic orbit—while certainly broadly present— are as yet highly inconclusive.

We do not understand why the nature of star formation has changed over the years. Why were globular clusters made for so long but no longer, and why does the disk now just manufacture the sparsely populated open clusters? What, if anything, is the role of dark matter in the evolution of the system? The Galaxy is a big, messy, chaotic assembly that we are only beginning to sort out.

Why, even, is the Galaxy a disk system, a normal spiral? Elliptical galaxies are very common. Why are we not one of them? What made one galaxy different from another? One possibility is different starting conditions. Ellipticals may have been rotating more slowly. They collapsed quicker, and star formation proceeded at a more rapid pace, using all the raw material and leaving the ellipticals with only old stars.

Collisions may also have played a role, even a dominant one. Within a dense cluster, galaxy interactions and collisions are quite likely. Such an event is not like the crash of two automobiles. Galaxies are empty affairs, and the stars within them will not bump together. Their interaction is gravitational. When two systems approach, they will raise huge tides in each other, distorting and disturbing otherwise regular orbits. They can in fact dissipate one another's kinetic energy and actually merge. The orbits of the stars then become chaotic and the combined system plumps up into an elliptical. Much of the matter in a cluster is actually stripped away and falls into the center, where it is consumed by a growing giant elliptical like M 87. Dwarf ellipticals may be tidal remnants. This picture explains why there are few spirals in dense clusters: They have to be more isolated to survive. We see our Milky Way at night only because we never encountered our neighbor M 31.

Why are we not a barred spiral? There is strong theoretical evidence from computer simulations that disk galaxies should develop a bar. In fact, we may well have one, but since we are inside our own galaxy, it is difficult to see. Bars seem to be suppressed by halos and bulges, possibly explaining why our bar (if indeed it exists) is not more prominent.

The next stage in development and evolution, the current formation of stars, is less mysterious. We feel we know what is actually going on.

THE INTERSTELLAR MEDIUM

Stars come from the matter that lies in the spaces between them, its existence made vividly clear by the diffuse nebulae and by the structure of the Milky Way itself, with its clouds of obscuring dust. These phenomena, however, represent but the simplest features of an enormously complex system only recently revealed to us by an advancing technology that allows observations by radio telescopes operating at a variety of frequencies and by ground- and space-based infrared detectors. Only when the intricate details of the interstellar medium are understood can we comprehend the complicated mechanisms by which stars are created.

The existence of a general interstellar gas pervading the galactic disk was first revealed in 1904, when astronomers noted the K line of Ca II in

NGC 4038 and 4039 collide; the tidal forces send long streams of matter into space. The collision foments vigorous star formation, seen at right.

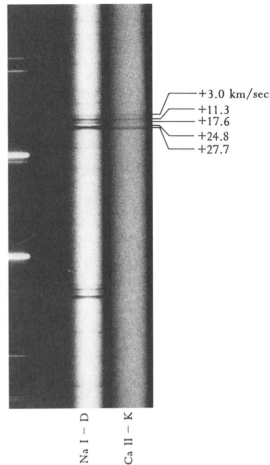

+3.0 km/sec
+11.3
+17.6
+24.8
+27.7

Na I – D   Ca II – K

These spectrograms show signatures of five separate clouds of interstellar matter along the line of sight to the star. The moving gas imprints absorptions of both the K line and the twin D lines.

absorption in the spectrum of Delta Orionis. This star is a spectroscopic binary whose stellar absorption features move back and forth over a 5.7-day period as a result of Doppler shifts. But the Ca II line just sits there unmoving, revealing that it is not a part of the star but must be produced in the space along the line of sight. Today we know of hundreds of these interstellar lines (particularly in the part of the ultraviolet recently made accessible by spacecraft), formed by a wide range of elements that includes neutral and ionized carbon, sodium, silicon, magnesium, zinc, nickel, and iron. Even the traces of some simple interstellar molecules—CH and CN, for example—are found in stellar spectra, a portent of the great discoveries of interstellar molecular chemistry to come. Quite clearly, interstellar gas is not confined to diffuse nebulae. The lines are commonly split by small Doppler shifts, showing that the matter is arranged in discrete clouds with densities of 10 or so atoms per cubic centimeter that are slowly moving (at a few km/s) relative to one another.

The insidious nature of the dust delayed its discovery. Before 1930, astronomers thought that except for the few obvious dark clouds, interstellar space was clear and transparent. That notion was shattered by Robert Trumpler of Lick Observatory in his research on open clusters, the kind found in the galactic disk. Distances determined on the basis of stellar brightness through main sequence fitting (see Chapter 3) were invariably larger than those found from the clusters' angular diameters based upon the assumption of common physical size. Moreover, the stars of the more distant clusters appeared redder than expected for their spectral classes. Suddenly it was clear that dust was everywhere in the galactic plane, dimming the light of the stars by about a magnitude for every thousand parsecs of distance and making the clusters appear farther away than they were. At about the same time, Edwin Hubble finally showed that the fuzzy "spiral nebulae" are actually distant galaxies, stellar systems similar to our own. It had long been known that they did not appear in the plane of the Milky Way. The obvious explanation of this "zone of avoidance" was the dust, which almost completely prevents us from seeing through the plane of the disk to the outside. The dust is distributed very irregularly, in great clumps and patches, the thickest of which, the Bok globules (after the Dutch-American astronomer Bart Bok, 1906–1983) are easily seen against the background of diffuse nebulae and the Milky Way.

The dimming or extinction of background starlight is caused by the scattering of photons rather than by absorption. The scattering efficiency increases in inverse proportion to wavelength, causing distant stars to appear anomalously red and demonstrating that the particles are typically well under a micron (a thousandth of a millimeter) in diameter. The dust is optically illuminated only when it lies in the neighborhood of a bright star, where it produces a reflection nebula. It is always found mixed in with the gas, the particles making about 1 percent of the total mass. The dust has a powerful impact on studies of the Universe because any distances based on inferred absolute magnitudes (or vice versa) require correction; fortunately, total extinction is related to the degree of reddening, which is usually measurable. The dust is so thick, however, that regions like the galactic center and sites of active star formation (the interiors of Bok globules, for example) are forever hidden from optical view.

Clearly, any picture of the Universe confined to the optical band of the electromagnetic spectrum must be very limited. A true understanding of the nature of interstellar space had to wait for developments in radio and infrared astronomy. These longer waves can penetrate the dust and provide information on low-temperature systems that have no optical effects.

The great initial step was taken in 1951 with the discovery by Edward Purcell and Harold Ewen of Harvard (based on a 1944 prediction by the Dutch astronomer Hendrik van de Hulst) that interstellar neutral hydrogen produces a powerful spectrum line at a wavelength of 21 centimeters. A magnetic interaction between the electron and the proton splits the ground energy state of the hydrogen atom into two very closely spaced hyperfine

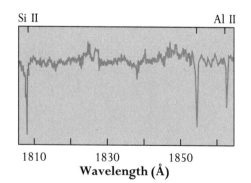

This spectral tracing shows ultraviolet absorption lines of interstellar Si II and Al II.

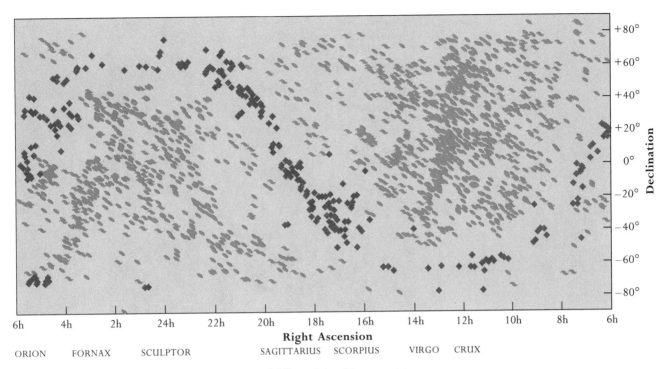

The plane of the Milky Way is shown by the distribution of diffuse nebulae (blue squares) in this map of the entire sky. The brighter galaxies (red ovals) completely avoid the plane because of an all-pervading sheet of dust.

levels. The energy is a little greater when the proton and electron are spinning in the same direction (parallel) than when they are spinning oppositely (antiparallel). The electrons are knocked into the parallel state by collisions with neighbors. When they spontaneously reverse direction, they emit photons at 21 cm. If the hydrogen happens to be silhouetted against a bright radio source, the 21-cm line can also be seen in absorption. Doppler shifts and radial velocities of discrete radiating or absorbing clouds are easily found, allowing the determination of the speed of galactic rotation as a function of distance from the center, as well as the estimation of the distances of the individual clouds. As a result, we see that the neutral hydrogen outlines spiral arms, and we can map them across the Galaxy.

But the 21-cm observations show only a small part of the picture. Beginning in the 1960s, radio astronomers began to find an abundance of interstellar molecules far in excess of that inferred from optical data. Mole-

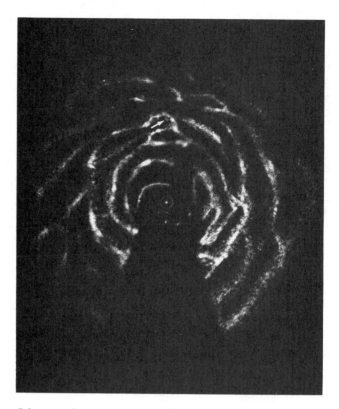

Galactic spiral structure as deduced from 21-cm line observations.

cules are free to rotate and vibrate in quantized energy states that split the electronic energy levels into fine structures. These are again further split into hyperfine levels. The transitions between a variety of closely spaced states produce low-energy photons in the radio spectrum. The first to be discovered were simple structures—the OH radical, ammonia, water, and carbon monoxide—several of them observable in circumstellar shells. Then came real astonishment, as rarer and more complex species were uncovered: formaldehyde, acetylene, methyl and ethyl alcohol, and (including some ions) nearly 80 more. The most intriguing—long organic chains of up to 13 atoms that cannot be produced in terrestrial laboratories—can be identified only by predicting their complex spectra theoretically.

The vast majority of the gas is simple molecular hydrogen, H_2, which is quite difficult to observe as it does not produce radio lines. However, we theoretically expect, and indeed observe, that the different kinds of molecules are linked together as one is created from another. Therefore, the location of the H_2 is traced out by powerful emissions of much less abundant carbon monoxide. Some molecular gas resides in globules, but most of it is found within giant molecular clouds with which the globules are frequently associated and whose dusty natures protect the fragile molecules from destructive stellar radiation. Some 6000 giant molecular clouds are known. Up to 100 pc across, they are the most massive discrete structures in the Galaxy, typically containing enough matter to create over 200,000 Suns. They are closely confined to the spiral arms of the Galaxy; indeed, it seems to be the spiral arms that make the clouds in the first place.

These graceful features are not permanent streams of stars, which would wrap up as the galaxy rotates. Instead, they are waves of density propagating through the interstellar material. They are initiated by gravitational perturbations possibly produced by a rotating bar or a nearby companion like the Large Magellanic Cloud. The gravitational disturbances pile up matter, producing more disturbances, and the waves then spread throughout the galaxy, developing a spiral pattern as the result of galactic rotation. Stars and clouds move in and out of them as they go by. The waves compress the interstellar gas, creating regions of higher density, and the clouds develop with the aid of their own gravity. These vast molecular assemblies are not the regions that produce the 21-cm emissions, which tend to lie along the outside edges of the true spiral arms. There is an evolutionary relation between the two that probably involves the way interstellar gas is consumed and transformed in star formation, but its nature is far from clear.

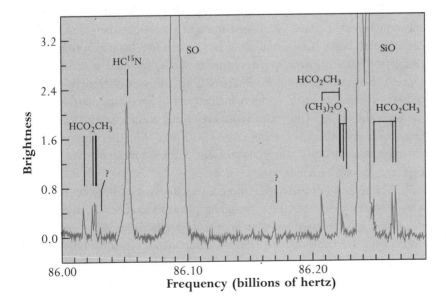

A small piece of the radio spectrum of the Orion Nebula at a frequency of 86 gigahertz (billion-cycles-per-second) shows emission lines of surprisingly complex molecules.

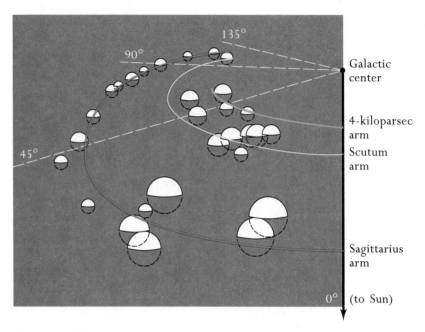

Molecular clouds lie along one of the inner spiral arms of the Galaxy.

The last step involves a more difficult spectral domain, the infrared. Advances in understanding the interstellar medium had to wait for improved detector technology and the ability to observe from space. Once observation became possible, the view was stunning. At longer wavelengths the cool, dusty clouds are no longer dark but glow brightly. Spectral absorption bands in heavily reddened starlight show that the grains of dust come in two basic varieties: silicates, and carbon in a form akin to graphite (recall the same division in circumstellar shells, outlined in Chapter 5). Optical chemical analysis of the interstellar gas shows it is strongly depleted in heavier elements, such as iron, which appear to have condensed from the gas onto the grains. The grains grow over time, accumulating atoms from interstellar space and becoming coated with ices. Much of the molecular chemistry takes place on the grain surfaces. After the molecules are formed, they are kicked back into space.

The infrared gives an even more intriguing view of interstellar chemistry at work. Several objects, including planetary, reflection, and diffuse nebulae, exhibit infrared emission lines that have been identified with polycyclic aromatic hydrocarbons (PAHs), components of soot built up from benzene rings. Alternatively, the spectral features may be produced by C_{60}, a soccer-ball shaped molecule named buckminsterfullerene after its structural resemblance to a geodesic dome and known familiarly as "bucky-ball." The deeper we look, the more we see. What are the most complex molecules that exist? Can we find the seeds of life itself?

Big clouds, little clouds, some dense, some tenuous, all netted within a thinner medium of neutral hydrogen and partially ionized gas. Occasionally, some of this gas comes in contact with a hot star, and the ghostly glow

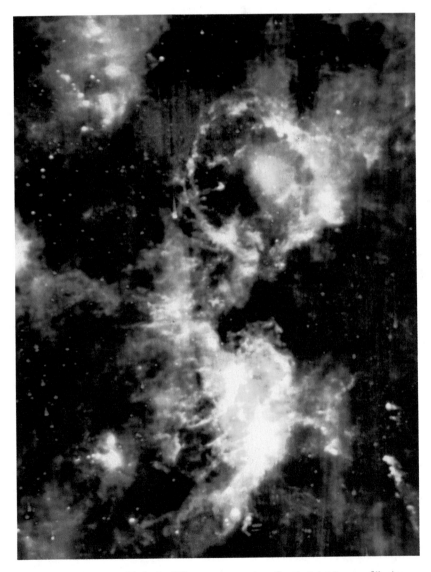

Dusty Orion takes on a completely different aspect in the infrared. A bright area fills the bottom half of the constellation; at the top, Lambda Ori is encircled by a huge, wind-blown bubble.

of a diffuse or reflection nebula marks its secret existence. Threading its way through this chaos, like tunnels in a giant sponge, is a hot gas, its temperatures in the hundreds of thousands of degrees, the result of overlapping bubbles created by supernovae.

INFANT STARS

We cannot formulate the processes by which stars develop without further evidence. Can we find newborn stars? If so, what are their characteristics?

Look at the dark clouds of the Milky Way in optical radiation and you see nothing, only a blank wall of dust. But infrared radiation can penetrate the obscuring haze, and we see that the clouds, from the small Bok globules to the larger complexes that sprawl over constellation boundaries, are filled with stars. We immediately surmise that these stars are very young and that the clouds are their birthplaces. That supposition is powerfully supported by the massive but short-lived O stars that are always associated with interstellar clouds, which they illuminate in great diffuse nebulae. Somehow, the gas and dust conspire to produce the thousands of mature lights that fill the evening sky—and the one that fills the day.

Proximity alone, however, does not prove that the clouds are stellar nurseries, nor does it tell us anything about the birth process itself. We need an observational progression of formation from the very beginning right onto the main sequence. And we have it. Sprinkled across the faces of the vast, dark clouds that sprawl through Orion, Scorpius–Ophiuchus, and Taurus–Auriga are myriad odd variables named after their prototype, T Tauri, which has been known for 50 years. Rather like the O stars, they are loosely organized into gravitationally unbound systems called T associations, showing that they had to be created together. They are typically G and K stars that exhibit considerable instability; T Tauri itself, for example, is normally about magnitude 11, but can erratically brighten to 10th or fade to 14th. Their youth is revealed by their spectra, as they exhibit strong absorption lines of lithium. This element is easily destroyed by nuclear reactions at temperatures far below those required for fusion. The lithium lines in the 5-billion-year-old Sun, for example, are quite weak. As convection circulates the solar gases downward to higher temperatures, the lithium in the envelope gradually diminishes. But the T Tauri stars have a full complement, comparable to that found in interstellar space.

Emission lines in T Tauri spectra disclose that the stars are involved with circumstellar gas. Many have P Cygni lines that reveal gas outflows,

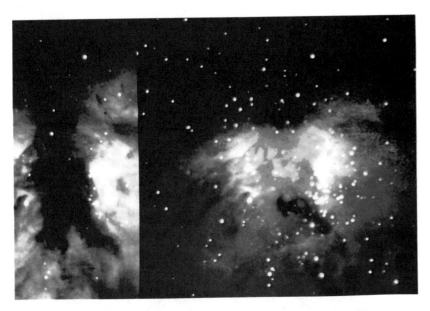

The optical view of this dark cloud (left) shows no hint of the activity within. However, infrared radiation (right) punches through the dusty fog, allowing us to see masses of stars.

but we occasionally also see *inverse* P Cyg features that demonstrate matter flowing *onto* the star. These stars are so young that they are still growing, accreting mass from their surroundings. One, DR Tauri, can show both infall and outflow at the same time! As in their evolved cousins, accretion seems to take place from a disk that surrounds the new star. The highly unstable nature of the accretion process accounts for at least some of the erratic variations that these stars undergo. Occasionally, the infall rate will suddenly increase, and the star will brighten by several magnitudes.

Long- and short-wave spectra reveal more surprises. The stars are much brighter in the infrared than they should be for their spectral types, the enhancement coming from clouds (or a disk) of circumstellar dust warmed by the stellar radiation. The ultraviolet is anomalously bright as well, a result of rapidly infalling gas and intense chromospheric activity that can also generate powerful X-ray flares. Since stellar activity steadily declines as a result of magnetic braking, the strong chromospheres again demonstrate how truly young these stars are.

The roles of the T Tauri stars in the flow of stellar evolution are clear when they are placed on the HR diagram, where they lie well above and to

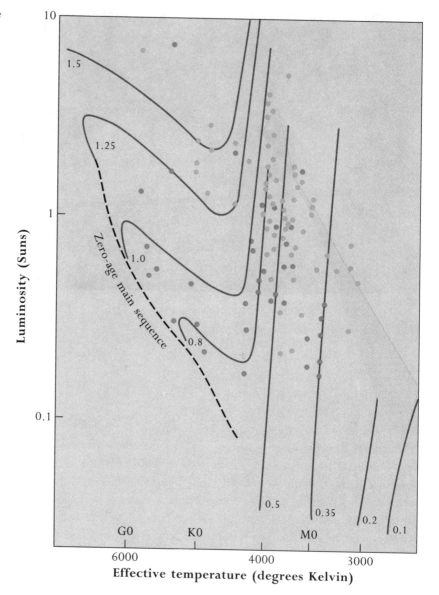

The T Tauri stars fall up and to the right of the main sequence (dashed line). The solid curves show theoretical tracks for stars of different mass. On the vertical portions, the stars are fully convective. The convection in the core shuts down as they evolve to the left and thermonuclear burning begins to stabilize. The shaded region shows the theoretical birthline, the dividing line between collapsing clouds of interstellar material and real stars. The stars fit their expected positions. Naked T Tauri stars are plotted as red dots; as expected, they are further along in development.

the right of the main sequence. Since they are demonstrably young, they must be evolving onto it rather than away from it. Further proof of the direction of evolution is that the so-called naked T Tauri stars, those that display only a vestige of the activity of this class, are closer to the zero-age main sequence than are the others. They have shed or used up much of their circumstellar disks and shrouds, enabling us to see through to the stars themselves with little interference. The whole collection falls right into the region predicted by theory for masses between about a third and double solar. Farther up the main sequence we see hotter-emission-line stars up to about 8 solar masses behaving the same way. Given time, they will all move to the left along their evolutionary tracks and become good, quiet citizens of the main sequence like our own Sun—which some 5 billion years ago must surely have been one of these unruly infants.

DISKS AND JETS

In the 1950s, George Herbig of Lick Observatory and Guilliermo Haro of the University of Mexico independently found a few small blobs of bright matter among the dark clouds of Orion. These Herbig–Haro (HH) objects show considerable structure, consisting of several individual knots. They are everywhere that we find dark clouds and T Tauri stars. The knots display emission lines but have no stellar cores nor obvious means of illumination. Even odder, their structures are changeable over a few years, a new knot appearing where none had been before. For a time, astronomers thought the objects were new stars undergoing gravitational collapse. Instead, these by-products of star formation provide powerful clues to the process.

HH objects are commonly paired, as seen in the photograph of HH 34. When we look closely enough, with deep optical or infrared imaging, we invariably find a star between the two. Some of the central sources are in fact T Tauri stars, so the phenomenon is clearly linked with stellar youth. Emerging from the star are oppositely directed twin jets of gas that are aimed right at the HH objects. These blobs are not new stars at all, but condensations of the interstellar medium that have been rammed and compressed by a bipolar flow of matter speeding along at several hundred kilometers per second. Over time, we can even see the HH objects being pushed radially outward under the force of the blow. Their illumination comes from the pounding shock wave that is set up as the jet comes to a screaming halt.

The narrowness of the twin beams is remarkable. Something within is tightly collimating them. Most likely, the central stars are surrounded by thick circumstellar disks, the same ones suspected of enclosing and feeding matter to the T Tauri stars, which display their own evidence for mass outflow. The two groups of objects are thus bound even more tightly. Evidently mass loss is a natural part of mass accretion. As matter falls into

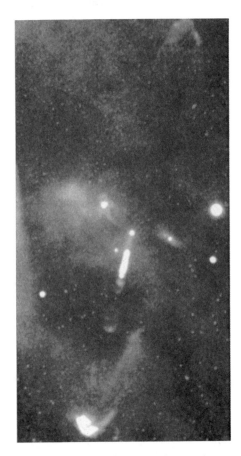

HH 34 is a complex of two HH objects with a new star in between. The star is spraying matter in opposing jets that produce shock waves in the surrounding medium.

Millimeter-wave observations of a bipolar object called L 1551 show flows of carbon monoxide moving toward (blue) and away (red) from us at about 10 km/sec. An energetic new star is developing in the middle (+), ejecting matter through the poles of a disk.

the star from the thick surrounding ring, some of it is hurled away, but it can escape only through the poles of the disk, where there is little or no mass. Exactly the same phenomenon was encountered for active galaxies with black hole cores and, in Chapter 6, for the one-time supernova and black hole candidate SS 433. In those cases the matter enters the disk from an evolving companion or from disrupted stars; with HH objects, it is acquired directly from the interstellar medium. The evidence that it occurs is overwhelming.

If we pull back from HH 34 to look at its environs, our view encompasses more and more of these fascinating objects. Several stars are being formed at the same time from the parental cloud, generating what will eventually be a T association. More remarkably, the jets are nearly all aligned! Something, most likely a broad magnetic field, has choreographed the spins of the interstellar clouds that collapse to create first the HH objects and then the T Tauri stars. The existence of such magnetic fields within the Galaxy is demonstrated by the partial polarization of obscured starlight (perfectly polarized light waves oscillate in a single plane), showing

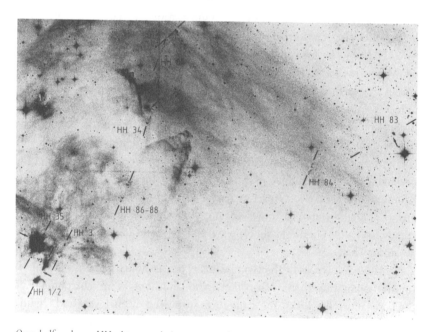

Over half a dozen HH objects and their associated twin jets surround HH 34, almost all with their axes aligned.

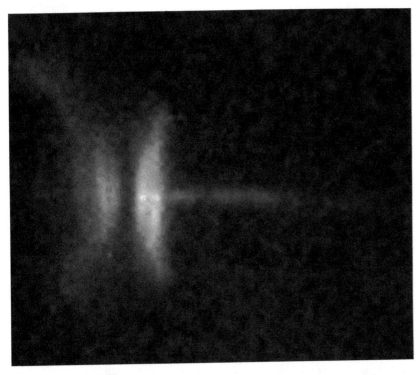

The Hubble Space Telescope shows a dark disk set perpendicular to two emerging flows. The edges of the disk are illuminated by the invisible star within.

that the elongated grains that make up the clouds and filter the light are magnetically constrained to lie in a common direction.

Evidence for bipolar flows is not confined to the obvious HH objects. Extending far beyond these optical shocks can be long streams of matter traceable with radio radiation from the CO molecule. Doppler shifts show that the CO is streaming away from a central source.

Despite this ample evidence, however, the collimating disks have so far in the discussion only been hypothesized. They are indeed there, and they can be seen. The ammonia molecule is a good tracer of dense matter. Radio observations show disks of ammonia radiation set perpendicular to the streams of CO emission. *Hubble* has even shown one directly.

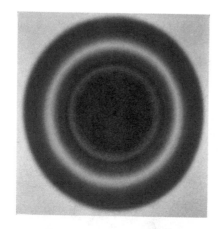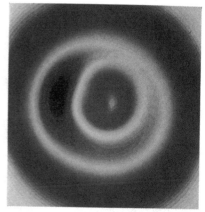

FORMATION

Star formation is as complex as stellar evolution. There is no single scenario, no one simple sequence that takes stars from their conception to their births as hydrogen-burning dwarfs. Some clouds generate high-mass stars, associations, even tight clumps of O stars, whereas others specialize in the lower main sequence. Whole clusters can be born with all stellar masses aboard, and so can isolated multiples, pairs, perhaps even singles. We do not yet understand why some clouds behave differently from others, nor are the paths of star formation all mapped out in detail. But the broad outline is becoming abundantly clear. So now we gather the evidence and place it in a context of theory.

The first step in star formation is the production of these clouds. The chief triggering agents seem to be the density waves that create the Galaxy's spiral structure. The clouds, once formed, are held together by their own gravity but are kept from collapsing by their internal turbulence, rotation, and quite likely by weak but extensive magnetic fields. However, here and there, self-gravity will win out, and the cold gas and dust will begin to contract into more compact zones called dense cores. At least some of these are likely to be triggered by blast waves from supernova explosions that provide additional compression, star death initiating star birth.

The big problem for a developing blob is its angular momentum. As the blob contracts, it must spin faster and faster. If nothing intervenes, the nascent star will tear itself apart long before it has a chance to be born. There are two chief ways of disposing of the excess angular momentum. The magnetic fields that thread through the clump of matter are tied to the surrounding gas. They can transfer the angular momentum outward, and thereby apply a brake, or the condensing mass can divide into two or more

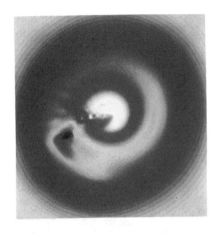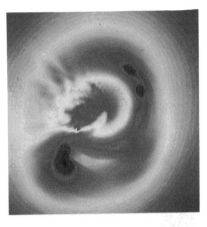

Supercomputer modeling shows how a cloud of contracting matter first divides in two; each clump then further fragments, and the result is a double-double star, a common configuration.

parts, the angular momentum going into orbital rather than rotational motion. Presumably, cloud fragmentation is the reason we see so many double and multiple stars: They are a natural by-product of the star-forming process.

Follow along now the development of stars like the Sun. The dense cores begin to collapse from the center outward, accreting more matter as time proceeds. At their centers they are developing protostars that over a few hundred thousand years will build to real stellar proportions. Growing protostars will become hot and very bright as a result of the conversion of gravitational energy into radiation. Once they become hot enough at their centers (about a million degrees Kelvin) the deuterium invested by the Big Bang begins to fuse into helium, so the protostars swell. The stars become fully convective, so that the deuterium falling in from the surrounding dense cores can be swept into the protostars and burned, which maintains their sizes. At the same time, the accreting matter is spinning out into disks as a result of the conservation of angular momentum. With radio telescopes, we now detect bipolar flows that emanate from the poles in the disks. The protostars then cross the "birthline" and now become visible as producers of Herbig–Haro objects and finally as T Tauri stars.

The winds of the protostars eventually develop winds, which terminate the infall and begin to clear away the surrounding opaque hazes. As the stars slowly move along their evolutionary tracks the deuterium fuel supply and thus deuterium burning diminishes and then finally stops. The stars then move toward the zero age main sequence. Eventually, the intense surface activity generated from the accretion disks diminishes and the naked T Tauri stars emerge. These pre–main sequence stars finally quiet down, become hot enough to fuse common hydrogen in their cores by the proton-proton chain, and settle into life as ordinary dwarfs. The whole process has taken a few tens of millions of years. The newly minted stars will now leave their birthplaces to wander free in space and to live serenely for billions of years, their origins long lost in time.

The higher-mass stars—up to about 8 solar masses—behave similarly. Some of the phenomena associated with star birth reach their acmes among the high-mass O stars, which not only produce bipolar flows but

also radiate with such intensity that they can sweep great bubbles in the interstellar medium, aiding in its compression. The brilliant O and B stars live such short lives that they never move very far before they expire. They therefore act as markers of the spiral arms and show us where stellar birth takes place. Without them, the arms, which are only enough to get the birth process under way, would effectively disappear from optical view.

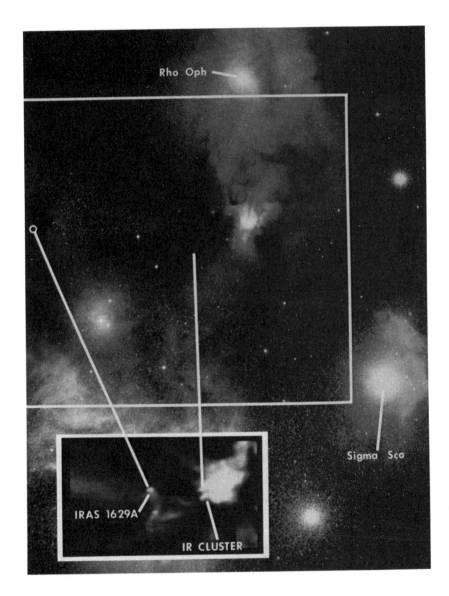

Spectacular bright and dark clouds fill the Milky Way on the Scorpius-Ophiuchus border in the optical photo at far left. The inset in the optical picture (near left bottom), taken in the infrared by IRAS, shows both a new cluster and a forming star in the Ophiuchus dark cloud. Radio emission from carbon monosulfide shows gas still accreting onto the star. Eventually the amount of dark matter will diminish, and new stars will help sweep the remainder away. In a few million years these stars will be seen optically.

PLANETS

The story of the disks does not end with stellar birth. Examine our own Sun. All its planets orbit in nearly the same plane and in the same direction. With three exceptions (Venus, Uranus, and Pluto) they also rotate in this direction, their axes more or less perpendicular to their orbits. Furthermore, this plane coincides with the solar rotational equator. Everything indicates that the bodies of the planetary system formed at the same time as the Sun and that they were born from a spinning disk. What disk could it be other than the one that we have seen develop all the way from the molecular clouds through to the T Tauri stars?

As the Sun ripened in the center, the dust within the circumstellar disk—the solar nebula—began to accumulate, first into substantial grains and then gradually into larger bodies. The immediate result was trillions of primitive planetesimals, each a few kilometers across. The compositions of the planetesimals depended on their distances from the Sun. In the inner part of the rotating cloud, where the solar heat was great, only refractory elements like metals and silicon could condense to produce silicates—rock. Beyond about 4 AU, the temperature was low enough for water in the surrounding nebula to turn to ice and be incorporated into, or to coat, the growing bodies. The planetesimals continued to collide and grow. Some eventually became large enough to dominate their neighborhoods and to sweep up the local materials as they turned gradually into real planets.

The violence of the early solar system must have been awesome, as bodies hundreds or even thousands of kilometers across hurtled into one another. The heat generated by the constant bombardment was so great that the new planets melted. The constituents of the dust separated much like the stuff in a blast furnace; metals settled to the center, and the slag—silicates—floated to the top. When cool, the planetary bodies were differentiated, with iron cores and rocky mantles and crusts.

In the inner solar system, the Sun blew away the light elements, the hydrogen and helium of the solar nebula, leaving the terrestrial planets—Mercury, Venus, Earth, and Mars—behind. In the outer part of the system, where the Sun had less influence, the solar nebula survived for a while, and much of it was accumulated by what are now the giant Jovian planets, Jupiter, Saturn, Uranus, and Neptune. Each of the giant planets had accumulated circulating disks as well, out of which their satellites were created. Our own Moon seems to have been made in a giant collision between the Earth and another primitive body.

The new planets mixed and swept up the remaining debris. After the smaller rocky bodies—the inner planets and all the satellites—cooled, they were subject to heavy bombardment by remaining planetesimals and broken detritus for another billion years, resulting in the heavy cratering still so evident on the airless surfaces of the Moon and Mercury. The water that had been cooked from the early Earth by the Sun and the heat of formation was then returned, ultimately providing our rains and our oceans. Even our carbon may have been accreted in this way.

Some ancient planetesimals still survive today. In between Mars and Jupiter are thousands of small rocky and icy bodies called asteroids. Jupiter kept them in such a state of agitation that repeated collisions among them broke them down and never allowed them to accumulate fully. They are always being tossed into orbits that cross that of the Earth, and occasionally they collide with us. A big one will produce a crater; a small one may lie in a field where it will someday be recognized as a meteorite to be studied as a sample of the days when the solar system was young. Go to a museum and look at a piece of truly ancient history, a seed of the Earth itself.

Trillions of icy planetesimals still linger in the remnant of the ancient disk far beyond the most distant planet. More were ejected by the planets into the great Oort cloud (after the Dutch astronomer Jan Oort) that surrounds the solar system perhaps halfway to the nearest star. On occasion, this cloud is stirred by passing stars and interstellar clouds, and a few of these bodies are sent into elliptical orbits toward the Sun. As they pass the orbit of Jupiter they heat and the ice volatilizes, carrying with it a variety of chemicals and dust and wrapping the ball of icy rock in a cloud. The solar wind, radiation, and magnetic field combine to push the gas into a variety of long, streaming tails to create the wonder that is a great comet.

This picture strongly suggests that planets are a common by-product of star formation. If our Sun has them, so should many of the nearby stars. Are they there? In 1984 the *IRAS* satellite detected the unmistakable infrared signature of warmed dust surrounding Vega, the most ordinary of A dwarfs. Dust has also been detected around Fomalhaut and others. Only a year later, astronomers at the Carnegie Institution's Las Campanas Observatory in Chile had more dramatic evidence, an actual image of a dusty disk around the A5 dwarf Beta Pictoris. The disk, seen edge-on, extends 400 AU from the star, far beyond the dimensions of the planetary system. The dust contains silicates, and there is evidence for particles considerably larger than those of interstellar space; moreover, the gas of the disk appears to contain large lumps that are spiraling in toward the star. Perhaps the material in the disk is creating a set of planets. Planets may even already exist within the inner segment of the spinning debris.

But a dusty disk is still circumstantial. Where are the planets themselves? Finding them is a daunting observational task. If Alpha Centauri had

Beta Pictoris is surrounded by a disk of dust, seen nearly edge-on, to a distance of 400 AU from the star. Recent HST spectroscopy has revealed lumps of gas, suggesting the model at right.

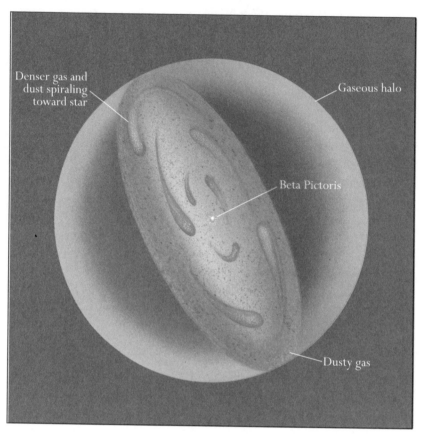

Denser gas and
dust spiraling
toward star

Gaseous halo

Beta Pictoris

Dusty gas

a planet like Jupiter, it would be about 22d magnitude, by itself an easy target. But it would also be only 4 seconds of arc from a brilliant star of magnitude zero, and thus totally lost in its glare. With our present techniques, we have little chance of seeing the planets directly. Yet there are other means of detection. An orbiting planet could cause the proper motion vector of a star to be noticeably deflected; that is, we are able to see a nearby star wobble back and forth as a result of gravitational effects. Such measurements are marginal for lower-mass stars and are impossible for bodies like the Sun, which would be diverted by only the extent of their own diameters. A much more sensitive test is to examine stellar radial velocities for periodic variations. The best modern technology can measure such deviations to within only a few meters per second. Several stars show such changes, suggesting planets that may be comparable to Jupiter's mass. More surprising are planets comparable to the Earth that have been found to be orbiting millisecond pulsars, apparently created from the debris of destroyed companions. They reveal themselves by subtle changes in the pulse arrival times. We are indeed approaching the point at which we can detect—if not see—real planetary bodies orbiting other stars. With our developing detector technology, it may not be too long before we have them in sight.

LIFE

What is always in the back of our minds, of course, is not so much the question of the existence of planets, but of what may be on those planets. Life evolved on Earth a mere billion years after planetary birth, as if it could not wait to get started. If it came so easily here, could it not also have come easily elsewhere? If planet formation is a natural by-product of star formation, might not life be a natural by-product too? And if there is life, can intelligent species have developed? Are there beings like us elsewhere who may be pondering the same questions, or are we alone?

The mere existence of planets hardly confirms life. In our own solar system only one planet managed to develop it. The others seem sterile, at least as far as we have been able to determine. Even Mars, the most likely candidate, is apparently quite dead. We, however, were lucky. Our planet was formed in just the right place, where it is neither too hot nor too cold, and where liquid water can exist. Only 0.3 AU inside our orbit, Venus is an

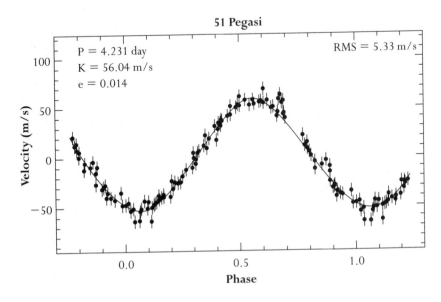

51 Pegasi

P = 4.231 day
K = 56.04 m/s
e = 0.014

RMS = 5.33 m/s

Velocity (m/s)

Phase

Fifth-magnitude 51 Pegasi wobbles about its center of gravity. Its radial velocity changes by a few tens of meters per second over a 4.2 day period, suggesting a planet only about half the mass of Jupiter.

uninhabitable oven. At the distance of Mars, conditions seem too cold, although there is ample evidence that water once flowed there.

Speculation about life, including intelligent life, elsewhere in space involves an equation, initially invented by American astronomer Frank Drake, that involves a combination of individual probabilities. What fraction of stars have planets? How many of these have them in a habitable zone, that is, one with liquid water? What fraction of them will actually develop life, and how frequently will these life-forms evolve into intelligent civilizations that may be able to communicate with us? Only the first question seems now to be addressable, as we are reasonably sure that at least a fraction of stars—perhaps all single ones, anyway—have planetary families. The question of life development involves such conjecture that we can obtain almost any answer we want. We have only one known case. The probabilities are boosted only by the staggering number of stars in the Galaxy. Even if the intrinsic probability of life is very low, there may still be many external civilizations; or there may still be just one.

In spite of the uncertainty, there is at least a possibility that somewhere out there other curious beings like us exist. Are they looking for us? Do we look for them? How might we make contact? The subject has long been taken seriously. However, it does *not* involve unidentified flying objects. UFOs are misperceived natural phenomena (most often sightings of Venus), delusions, or hoaxes. Interstellar space travel would involve speeds a good fraction that of light in order to cover the great distances involved, and the energy budgets for such ventures may place them entirely outside the realm of possibility.

The "search for extraterrestrial intelligence," or SETI, does not involve any suggestion of direct contact, but communication by radio. For the past several decades, the Earth has been radio-bright, beaming countless radio and television broadcasts into the cosmos. We have announced ourselves by a bubble of electromagnetic radiation that is now approaching 100 light-years (33 parsecs) in radius and encompasses thousands of stars. Maybe we can catch someone else's bubble. Or maybe they are deliberately sending messages in hope of finding company.

The search is daunting, the reward scientifically and philosophically beyond measure. How do we begin? The sky and the radio spectrum are both enormous. The first probe was undertaken in 1960 by Drake under Project Ozma with an examination of two nearby solar-type stars, Epsilon Eridani (which may have a planet) and Tau Ceti. He scanned radio frequencies near 21 cm because the powerful nature of the neutral hydrogen line makes it an obvious reference point; he found nothing. Modern SETI instruments are capable of observing simultaneously in a million radio channels and of looking vastly deeper into space. They, too, have found nothing.

So far, SETI is a science with one datum: ourselves. Is anyone else out there? Fifty years ago, we were ignorant of how the shining lights of the nighttime sky were made. Now we know; or at least we are convinced we are on the right track. What will we know a century hence? Will we have found planets and galactic company? The search may go on for as long as we care to wonder about the stars, their natures, their origins, and their fates.

A million stars appear in a segment of the Milky Way. Ours would be no more than a dot hiding an extraordinary treasure . . .

EPILOGUE

We are almost at the end of the journey. These seven chapters have outlined our centuries-old wanderings among the stars. We have looked at the ancient stories; the beginnings of knowledge; stellar development, death, and, finally, glorious birth. But there is still a perspective to gain.

Expand your view and look at the Galaxy as a vast recycling engine. Stars come dripping from the fonts of interstellar space, where they are created out of the loose gas and dust. As they age, they pump enriched matter back into the wellsprings of creation. Slowly the whole galactic disk becomes richer in heavy elements, as do the new stellar generations.

. . . a blue ball inhabited by beings who can turn their vision back into the depths of the sky.

Hiding themselves in dusty cocoons, the great evolved stars blow the tiny grains of silicon and carbon into space, where they grow fat on the cold ambient gas and aid in the construction of complex molecules that will make the icy, dense stellar seedbeds. Supernovae explode, and their blast-waves help compress the interstellar brew, thus furthering the process. Out of the old comes the new. Each generation assists in the creation of the next.

Planets form around a new Sun, accumulated from the dust left in the remnants of formation. Solar heat drives the lighter-weight materials away, leaving the denser behind. The Earth is the distillate of the heavens, each atom beyond helium produced by a dying star somewhere in the vastness of space. The recycling engine may operate on a deep and personal level: Perhaps a bit of the dust that helped create us came from the destruction of other planets that once encircled other stars.

We do not know how life began, where the initial building blocks came from. The water that made it possible was likely brought aboard shortly after the Earth was formed, when the countless remaining icy and rocky planetesimals rained down upon it during the time of the great bombardment. It may even be possible that some water originally drifted into our birth cloud from evolving giant stars that had boiled away their families of comets. The impacts surely also delivered complex molecules made in the chemical factory of interstellar space. We have no idea what substances can be built over the billions of years available; but amino acids have been found in meteorites. Perhaps—and only perhaps, as our ignorance is deep indeed—the basic molecular structure of life came from out there, too. If so, not only did stars and planets grow from the black clouds of the Milky Way, so did we.

The mighty stars, the vastness of space, the incomprehensible distances, the aeons of time since the beginning can inspire a feeling of insignificance. We seem so small compared to our surroundings. But look at it from a different point of view. In a sense, we—and perhaps others out there as well—are at the focus of it all. However life developed, it has taken the entire Galaxy and its evolution to make us. More importantly, we can look back into the darkness, at the hosts of stars swirling about our heads, and can *think* about them. It is we who have the real power. We can comprehend the Universe; we can appreciate its order and its beauty. We stand beneath the sky and can know the stars.

APPENDIX 1 THE ANCIENT CONSTELLATIONS

NAME	MEANING	α(h)	$\delta°$	GENITIVE	ABBREVIATION
		The Zodiac			
Aries	ram	3	+20	Arietis	Ari
Taurus	bull	5	+20	Tauri	Tau
Gemini	twins	7	+20	Geminorum	Gem
Cancer	crab	8.5	+15	Cancri	Cnc
Leo	lion	11	+15	Leonis	Leo
Virgo	virgin	13	0	Virginis	Vir
Libra[1]	scales	15	−15	Librae	Lib
Scorpius	scorpion	17	−30	Scorpii	Sco
Sagittarius[2]	archer	19	−25	Sagittarii	Sgr
Capricornus[3]	water-goat	21	−20	Capricornii	Cap
Aquarius[3]	waterman	22	−10	Aquarii	Aqr
Pisces[3]	fishes	1	+10	Piscium	Psc
		Ursa Major			
Ursa Major	larger bear	11	+60	Ursae Majoris	UMa
Ursa Minor[4]	smaller bear	16	+80	Ursae Minoris	Umi
Boötes	proper name; herdsman, wagoner	15	+30	Boötis	Boo
		Orion			
Orion	proper name; hunter, giant	6	0	Orionis	Ori
Canis Major	larger dog	7	−20	Canis Majoris	CMa
Canis Minor	smaller dog	8	+5	Canis Minoris	CMi
Lepus	hare	6	−20	Leporis	Lep
		Perseus			
Perseus	proper name; hero	3	+45	Persei	Per
Andromeda	proper name; princess	1	+40	Andromedae	And
Cassiopeia	proper name; queen	1	+60	Cassiopeiae	Cas
Cepheus	proper name; king	22	+65	Cephei	Cep
Cetus	whale	2	−10	Ceti	Cet
Pegasus	proper name; winged horse	23	+20	Pegasi	Peg
		Argo			
Carina[5]	keel	9	−60	Carinae	Car
Puppis[5]	stern	8	−30	Puppis	Pup

(continued)

(continued from previous page)

NAME	MEANING	α(h)	δ°	GENITIVE	ABBREVIATION
Vela[5]	sails	10	−45	Velorum	Vel
Hercules[6]	proper name; hero	17	+30	Herculis	Her
Hydra	water serpent	12	−25	Hydrae	Hya
Aries[7]	ram; here, the golden fleece	3	+20	Arietis	Ari

Centaurus

Centaurus[8]	centaur	13	−45	Centauri	Cen
Lupus	wolf	15	−45	Lupi	Lup
Ara	altar	17	−55	Arae	Ara

Ophiuchus[9]

Ophiuchus[9]	serpent-bearer	17	0	Ophiuchi	Oph
Serpens[9]	serpent	17	0	Serpentis	Ser

Single Constellations

Aquila	eagle	20	+15	Aquilae	Aql
Auriga	charioteer	6	+40	Aurigae	Aur
Corona Australis[10]	southern crown	19	+40	Coronae Australis	CrA
Corona Borealis[11]	northern crown	16	+30	Coronae Borealis	CrB
Corvus[12]	crow, raven	12	−20	Corvi	Crv
Crater	cup	11	−15	Crateris	Crt
Cygnus	swan	21	+40	Cygni	Cyg
Delphinus[3]	dolphin	21	+10	Delphini	Del
Draco[13]	dragon	15	+60	Draconis	Dra
Equuleus	little horse	21	+10	Equulei	Equ
Eridanus	proper name; river	4	−30	Eridani	Eri
Lyra[12]	lyre	19	+35	Lyrae	Lyr
Piscis Austrinus[3]	southern fish	22	−30	Piscis Austrini	PsA
Sagitta	arrow	20	+20	Sagittae	Sge
Triangulum	triangle	2	+30	Trianguli	Tri

[1] Originally the claws of Scorpius.

[2] Contains the galactic center.

[3] Constellations of the area of the sky known as the wet quarter for its many watery images.

[4] Contains the north celestial pole.

[5] Carina, Puppis, and Vela are modern subdivisions of the old constellation Argo and together make *one* of the ancient 48.

[6] One of the oldest constellations known.

[7] Also included in the Zodiac.

[8] Sometimes included in the Argo myth; the centaur Chiron was Jason's foster father.

[9] Ophiucus is identified with the physician Aesculapius, and Serpens with the caduceus.

[10] Sometimes considered as Sagittarius' crown.

[11] Ariadne's crown.

[12] Corvus was Orpheus' companion, Lyra his harp.

[13] Contains the north ecliptic pole.

APPENDIX 2   THE MODERN CONSTELLATIONS

NAME	MEANING	α(h)	δ°	GENITIVE	ABBREVIATION
Antlia	air pump	10	−35	Antliae	Ant
Apus	bee	16	−75	Apodis	Aps
Caelum	graving tool	5	−40	Caeli	Cae
Camelopardalis	giraffe	6	+70	Camelopardalis	Cam
Canes Venatici	hunting dogs	13	+40	Canum Venaticorum	CVn
Chamaeleon	chameleon	10	−80	Chamaeleontis	Cha
Circinus	compasses	15	−65	Circini	Cir
Columba	dove	6	−35	Columbae	Col
Coma Berenices[1]	Berenice's hair	13	+20	Comae Berenices	Com
Crux[2]	southern cross	12	−60	Crucis	Cru
Dorado[3]	swordfish	6	−55	Doradus	Dor
Fornax	furnace	3	−30	Fornacis	For
Grus	crane	22	−45	Gruis	Gru
Horologium	clock	3	−55	Horologii	Hor
Hydrus	water snake	2	−70	Hydri	Hyi
Indus	Indian	22	−70	Indi	Ind
Lacerta	lizard	22	+45	Lacertae	Lac
Leo Minor	smaller lion	10	+35	Leonis Minoris	LMi
Lynx	lynx	8	+45	Lyncis	Lyn
Microscopium	microscope	21	−40	Microscopii	Mic
Monoceros	unicorn	7	0	Monocerotis	Mon
Mensa	table	6	−75	Mensae	Men
Musca (Australis)[4]	(southern) fly	12	−70	Muscae	Mus
Norma	square	16	−50	Normae	Nor
Octans[5]	octant	—	−90	Octantis	Oct
Pavo	peacock	20	−70	Pavonis	Pav
Phoenix	phoenix	1	−50	Phoenicis	Phe
Pictor	easel	6	−55	Pictoris	Pic
Pyxis[6]	compass	9	−30	Pyxidis	Pyx
Reticulum	net	4	−60	Reticuli	Ret

(continued)

(continued from previous page)

NAME	MEANING	$\alpha(h)$	$\delta°$	GENITIVE	ABBREVIATION
Sculptor[7]	sculptor's workshop	1	−30	Sculptoris	Scl
Scutum[8]	shield	19	−10	Scuti	Sct
Sextans	sextant	10	0	Sextantis	Sex
Telescopium	telescope	19	−50	Telescopii	Tel
Triangulum Australe	southern triangle	16	−65	Trianguli Australis	TrA
Tucana[9]	toucan	0	−65	Tucanae	Tuc
Volans	flying fish	8	−70	Volantis	Vol
Vulpecula	fox	20	+25	Vulpeculae	Vul

[1] Star cluster with many old references, but not considered one of the ancient 48. Referred to by Eratosthenes as Ariadne's Hair; contains the north galactic pole.

[2] Originally a part of Centaurus.

[3] Contains the Large Magellanic Cloud and the south ecliptic pole.

[4] Originally named Musca Australis to distinguish it from Musca Borealis, the northern fly, which is now defunct; "Australis" is now dropped.

[5] Contains the south celestial pole.

[6] Grouped with the ancient constellation Argo.

[7] Originally named by Lacaille l'Atelier du Sculpteur (in Latin, Apparatus Sculptoris); now known simply as Sculptor. Contains the south galactic pole.

[8] Shield of the Polish hero John Sobieski.

[9] Contains the Small Magellanic Cloud.

NAME	MEANING[1]	GREEK-LETTER NAME	MAGNITUDE[2]	DISTANCE[3] (pc)	ABSOLUTE[3] MAGNITUDE	SPECTRAL[3] TYPE
Sirius	scorching (Gr)	α CMa	−1.46	2.64	1.43	A0 V
Canopus	proper name; pilot (Gr)	α Car	−0.72	96	−5.6	F0 II
Rigil Kentaurus	the Centaur's foot	α Cen A	−0.01	1.35	4.34	G2 V
		α Cen B	1.33	1.29	5.78	K1 V
Arcturus	bear-watcher (Gr)	α Boo	−0.04	11.3	−0.31	K1 III
Vega	swooping eagle	α Lyr	0.03	7.76	0.58	A0 V
Capella[4]	she-goat (Lat)	α Aur A	0.08	12.9	−0.47	G5 III
		α Aur B				G0 III
Rigel	foot of the Central One	β Ori	0.12	240	−7	B8 Ia
Procyon	before the dog (Gr)	β CMi	0.38	3.49	2.67	F5 IV
Achernar	end of the river	α Eri	0.46	44	−2.8	B3 V
Betelgeuse	armpit of the Central One (corr)	α Ori	0.50v	130	−5.1	M2 Ia
. . .	. . .	β Cen	0.61	160	−5.4	B1 III
Altair	flying eagle	α Aql	0.77	5.14	2.22	A7 V
Aldebaran	follower	α Tau	0.85	20.0	−0.66	K5 III
Antares	like Mars (Gr)	α Sco	0.96	190	−5.4	M1.5 Ib
Spica[5]	ear of wheat (Lat)	α Vir	0.98	80	−3.5	B2 V
. . .	. . .	α Cru A	1.58	98	−3.8	B0.5 IV
Pollux	proper name; twin (Lat)	β Gem	1.14	10.3	1.08	K0 III
Fomalhaut	fish's mouth	α PsA	1.16	7.69	1.73	A3 V
Deneb	tail	α Cyg	1.25	500	−7	A2 Ia
. . .	. . .	β Cru	1.25	110	−4.0	B0.5 III
Regulus	little king (Lat)	α Leo	1.35	24	−0.55	B7 V
Adhara	virgins	ϵ CMa	1.50	130	−4.1	B2 II
Castor[6]	proper name; twin (Lat)	α Gem	1.58	15.8	0.59	A1 V
. . .	. . .	γ Cru	1.63	36	−1.14	M3.5 III
Shaula	stinger	λ Sco	1.63	200	−5	B2 IV
Bellatrix	female warrior (Lat)	γ Ori	1.64	75	−2.7	B2 III
Elnath	butting one	β Tau	1.65	40	−1.4	B7 IV
Miaplacidus	uncertain; placidus = calm (Lat)	β Car	1.68	34	−1.0	A2 IV
Alnlinam	string of pearls; refers to Orion's belt	ϵ Ori	1.70	460	−6.6	B0 Ia

(continued)

(continued from previous page)

NAME	MEANING[1]	GREEK-LETTER NAME	MAGNITUDE[2]	DISTANCE[3] (pc)	ABSOLUTE[3] MAGNITUDE	SPECTRAL[3] TYPE
Al Nair	bright one in the fish's tail	α Gru	1.74	31	-0.7	B7 IV
Alioth	the bull; (corr)	ϵ UMa	1.77	25	-0.2	AO V
Regor	uncertain derivation	γ^2 Vel	1.78	260	-5	WC8 + 09 I
Dubhe	bear	α UMa	1.79	38	-1.1	K0 III
Mirfak	elbow	α Per	1.79	180	-4	F5 Ib
Wezen	weight	δ CMa	1.84	740	-7.5	F8 Ia
Alkaid	chief of mourners	η UMa	1.86	31	-0.6	B3 V
Avior	uncertain derivation	ϵ Car	1.86	190	-5	K3 III + B2 V
Menkalinen	shoulder of the rein holder	β Aur	1.90	25	-0.1	A2 IV

[1] Names are of Arabic derivation unless otherwise stated. "Lat" and "Gr" refer to Latin and Greek names; serious corruption is indicated by "corr." Principal source is P. Kunitzsch and T. Smart, *Short Guide to Modern Star Names and Their Derivations.* Weisbaden: Harrasowitz, 1986.

[2] Visual magnitudes, or *V* (see Chapter 2); "v" denotes variable magnitude.

[3] Distance in parsecs, absolute visual magnitude (M_v), and spectral types are discussed in Chapter 3. Distances under 400 pc are from Hipparcos parallax measurements; the others are derived from spectral class. These new values are not reflected in the HR diagram on p. 89.

[4] Visually unresolved binary; magnitudes are combined values.

[5] Binary with roughly equal components.

[6] Multiple star.

APPENDIX 4
STAR AND CONSTELLATION MAPS

The following six maps locate most of the constellations and brighter stars. The first shows the north polar region down to about 50° declination. The next four are seasonal equatorial maps that display stars between 60°N and 60°S declinations, and the last shows the south polar region. Declinations are shown along a central hour circle. Right ascensions are noted around the peripheries of the polar maps and along the celestial equator for the seasonal maps.

Stars, generally selected to indicate constellation positions and outlines, are shown through fourth magnitude, although their census is not complete. Fifth magnitude stars are included where they are important parts of their constellations. A few nonstellar objects—clusters, nebulae, and galaxies—are also noted.

The maps show the broad outline of the Milky Way; however, much of its intricate detail is omitted. The galactic equator (the mid-line of the Galaxy) is marked with galactic longitude starting at the galactic center in Sagittarius.

The dates along the edges of the polar maps and along the tops and bottoms of the equatorial maps indicate the appearance of the sky at approximately 8:30 P.M. (20^h30^m) local time. To use the north polar map, face north and rotate the map so that the current month appears at the top. In the southern hemisphere, use the south polar map similarly. The celestial pole will have an elevation in degrees equal to your latitude. To use the equatorial maps in the northern hemisphere, face south and line up the current month with the celestial meridian. To use them in the southern hemisphere, face north and turn them upside down. The equator point (the intersection between the celestial equator and the meridian) will have an elevation in degrees equal to 90° minus the latitude.

For each hour past $20^h 30^m$, shift or rotate the map one hour to the west (that is, align an hour circle that is one additional hour to the east). For every 2 hours past $20^h 30^m$, add one month to your current month. For example, if it is March 15 at $20^h 30^m$, you would align "March" on Map 4 with the celestial meridian. If it is $2^h 30^m$, set "June" (Map 5) on the meridian.

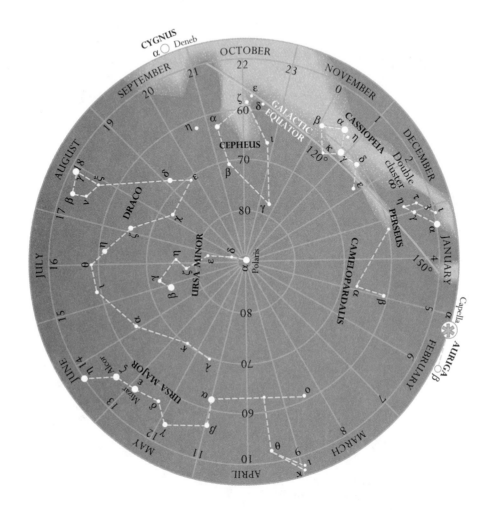

Map 1. The North Polar Constellations

SCALE OF MAGNITUDES

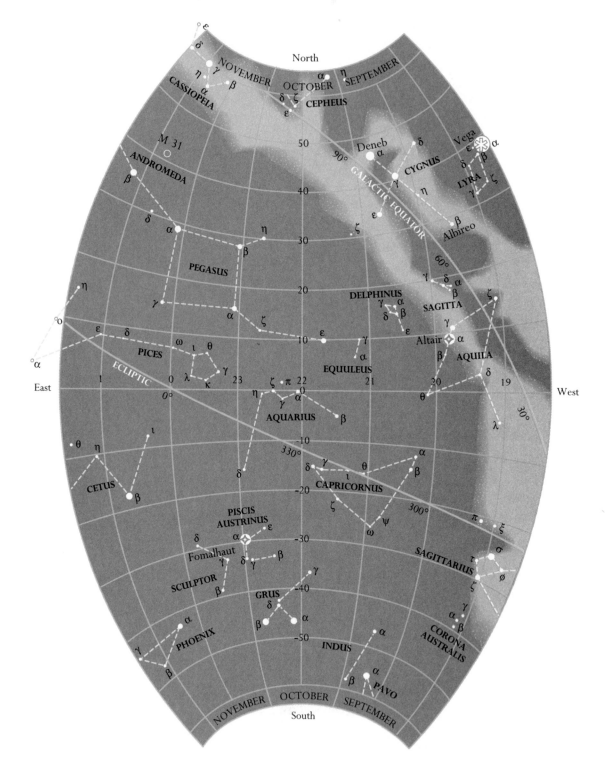

Map 2. The Constellations of Northern Autumn, Southern Spring

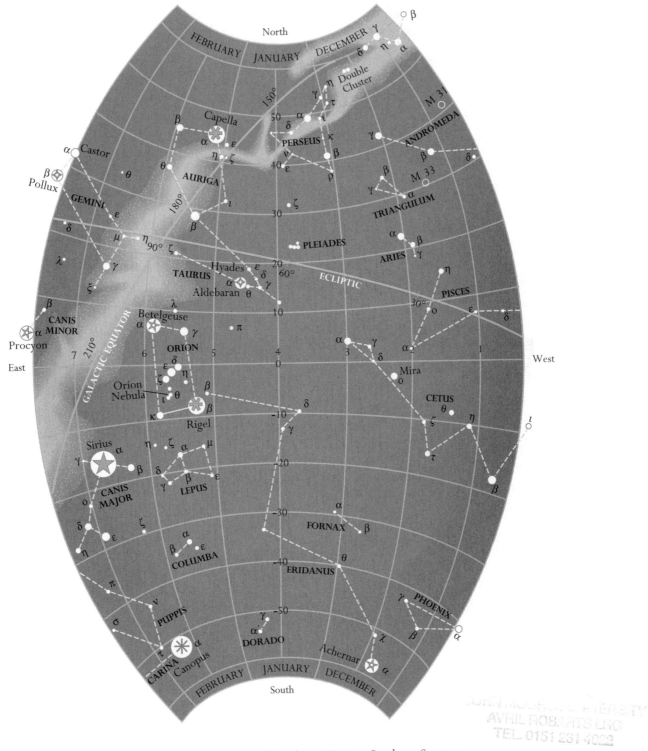

Map 3. The Constellations of Northern Winter, Southern Summer

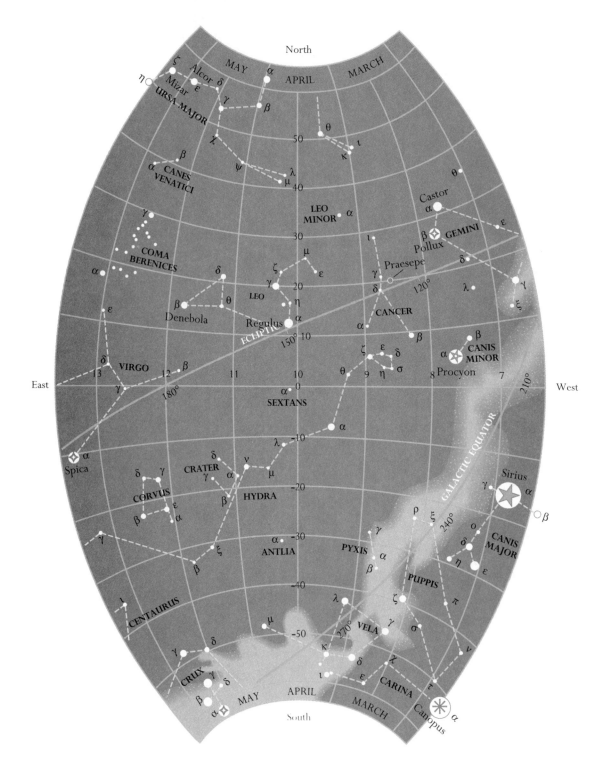

Map 4. The Constellations of Northern Spring, Southern Autumn

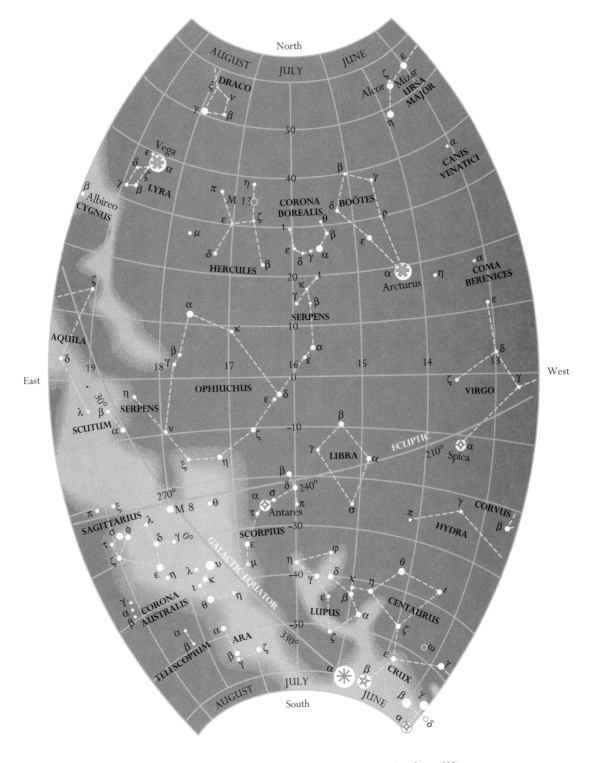

Map 5. The Constellations of Northern Summer, Southern Winter

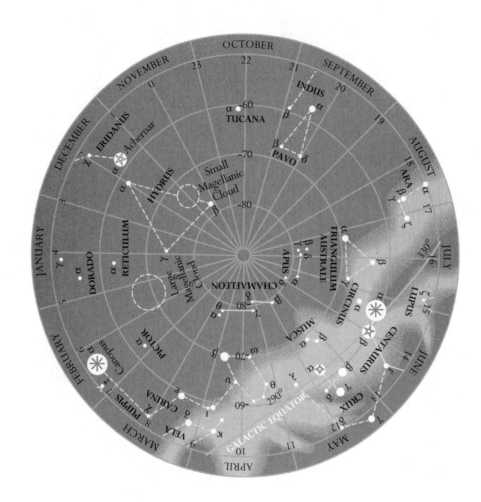

Map 6. The South Polar Constellations

APPENDIX 5
THE LARGEST TELESCOPES

REFRACTORS

40-inch (1-meter), Yerkes Observatory, University of Chicago, Williams Bay, Wisconsin, U.S.A.

36-inch (0.9-meter), Lick Observatory, University of California, Mount Hamilton, California, U.S.A.

REFLECTORS

16-meter Very Large Telescope optical array,[1, 2] European Southern Observatory, LaSilla, Chile

10-meter, Keck Telescope, California Institute of Technology and the University of California, Mauna Kea, Hawaii, U.S.A.[1]

10-meter, Keck II Telescope, California Institute of Technology and the University of California, Mauna Kea, Hawaii, U.S.A.

8-meter, National Optical Astronomy Observatories, Kitt Peak, Arizona, U.S.A., and Cerro Pachon, Chile[1]

6-meter, Caucasus Mountains, Soviet Union

5-meter, Hale Observatory, Palomar Mountain, California, U.S.A.

4.5-meter equivalent, Multiple Mirror Telescope, University of Arizona and the Smithsonian Center for Astrophysics, Mount Hopkins, Arizona, U.S.A.

4.2-meter, Royal Greenwich Observatory, Canary Islands

4-meter, National Optical Astronomy Observatories, Kitt Peak, Arizona, U.S.A., and Cerro Tololo Interamerican Observatory, Chile

RADIO[3]

43-kilometer Very Large Array (VLA) Interferometer, National Radio Astronomy Observatory, New Mexico, U.S.A.

20-km Westerbork Radio Synthesis Observatory Interferometer, Westerbork, Holland

305-meter fixed single dish, Arecibo Observatory, Puerto Rico

100-meter single dish, Max Planck Institute for Radio Astronomy, Bonn, Germany

100-meter single dish, National Radio Astronomy Observatory, Green Bank, West Virginia, U.S.A.[1]

---

[1] Planned or under construction. This table will date quickly.

[2] Array of four 8-meter telescopes with an effective light-gathering power of one 16-meter telescope.

[3] The largest "instrument" actually achieved was a very long baseline interferometer (VLBI) made with telescopes set 10,000 kilometers apart; a 5000-kilometer separation is routine.

"The Origin of the Milky Way," by *Tintoretto (ca. 1518–1594).*

Sources of Illustrations

Drawings by Ian Worpole, Chanticleer Graphics, and Vantage Art, Inc.

Frontispiece
Drawing by Ben Shahn, *Scientific American,* Sept. 1953

Prologue
"Starry Night over the Rhone" by Vincent van Gogh; Musée National, Paris. Giraudon/Art Resource

Page 4
The National Maritime Museum, Greenwich

Page 8
Roger Ressmeyer/Starlight

Pages 12–13
Mario Grassi

Page 14
Rick Olson

Page 15
Okiro Fujii/L'Astronomia

Page 16
University of Illinois at Urbana-Champaign Library

Page 17
J. Richard Hansen

Page 18
University of Illinois at Urbana-Champaign Library

Page 19
Dennis di Cicco

Page 20
top, Rudy Schild, Smithsonian Astrophysical Observatory
bottom, Jim Riffle, Cloudcroft, NM

Page 23
Greenwich Observatory

Page 27
Dennis di Cicco

Page 28
National Maritime Museum, Greenwich

Page 33
Art Resource

Page 34
Roger Ressmeyer/Starlight

Page 39
Bausch and Lomb International

Page 41
Art Resource

Page 43
SERC, reproduced from the ESO/SERC J Survey

Page 44
top, Helevius, Selenographia (Danzig 1647)
bottom, Roger Ressmeyer/Starlight

Page 46
Roger Ressmeyer/Starlight

Page 51
Pat Seitzer

Page 54
top, Roger Ressmeyer/Starlight
bottom, Roger Ressmeyer/Starlight

Page 55
Roger Ressmeyer/Starlight

Page 56
Space Telescope Science Institute, NASA

Page 57
NASA

Page 58
H. A. McAlister

Page 59
European Southern Observatory

Page 61
Roger Ressmeyer/Starlight

Page 62
Jim Riffle, Cloudcroft, NM

Page 68
Lowell Observatory

Page 69
SERC, reproduced from the ESO/SERC J Survey
insert, NOAO

Page 75
Deutsch Museum, Munchen

Page 77
The University of Michigan

Page 78
Whitin Observatory, Wellesley College

Page 79
The University of Michigan

Page 83
right, E. C. Olson, NOAO
left, From *Stars and Their Spectra* by J. B. Kaler, Cambridge University Press

Page 84
top, Landolt-Bornstein Tables, Springer, T. Schmidt-Kaler

bottom, From *Stars and Their Spectra* by J. B. Kaler, Cambridge University Press

Page 85
Princeton University Libraries

Page 86
Dennis di Cicco

Page 88
bottom, David Buscher, Chris Haniff, John Baldwin, and Peter Warner, Mullard Radio Astronomy Observatory, Cavendish Laboratory, Cambridge, UK
top, From *An Atlas of Representative Spectra* by Y. Yamashita, K. Nariai, and Y. Norimoto, University of Tokyo Press

Page 89
From *Stars and Their Spectra* by J. B. Kaler, Cambridge University Press

Page 90
Lick Observatory

Page 91
ROE/AAT Board 1985

Page 93
From *An Atlas of Open Cluster Colour-Magnitude Diagrams* by G. L. Hagen, David Dunlap Observatory

Page 95
left, Dennis di Cicco
left insert, Lowell Observatory
right, Yerkes Observatory

Page 96
E. C. Olson

Page 97
Lowell Observatory

Page 99
ROE/AAT Board 1984

Page 100
Anglo-Australian Telescope Board
1986

Page 102
left, Anglo-Australian Telescope
Board
right, NOAO

Page 103
Dudley Observatory Reports by
A. G. D. Philip, M. F. Cullen,
and R. E. White

Page 104
Leon Golub, IBM Research and
Smithsonian Astrophysical
Observatory

Page 107
National Solar Observatory/Sacra-
mento Peak

Page 108
E. C. Olson, Mt. Wilson and Las
Campanos Observatories

Page 109
top left, Dennis di Cicco
top right, National Solar Observa-
tory/Sacramento Peak
right bottom, Big Bear Solar Observatory, California Institute
of Technology

Page 110
Dennis di Cicco

Page 111
top, National Solar Observatory/
Sacramento Peak
bottom, J. A. Eddy, University
Corporation for Atmospheric
Research

Page 112
NOAO

Page 113
top, Big Bear Solar Observatory,
California Institute of Technology
bottom, J. B. Kaler, NOAO

Page 115
Naval Research Laboratory

Page 116
Jim Riffle, Cloudcroft, NM

Page 118
NASA

Page 120
Royal Astronomical Society,
London

Page 121
Harvard University Archives

Page 129
Physics Today

Page 134
E. C. Olson, NOAO

Page 135
Sproul Observatory

Page 136
Palomar Observatory

Page 137
S. Vogt, Lick Observatory

Page 138
Jack Marling

Page 143
Icko Iben, Jr., in *Annals of Physics*

Page 144
Icko Iben, Jr.

Page 149
Anglo-Australian Telescope Board
1977

Page 150
Allan Sandage

Page 151
J. E. Hesser, W. E. Harris,
W. E. VandenBerg,
D. A. Allwright, J. W. Shott,
and P. B. Stetson

Page 153
Icko Iben, Jr.

Page 154
American Association of Variable
Star Observers

Page 155
From *An Atlas of Stellar Spectra* by
W. W. Morgan, P. C.
Keenan, and E. Kellman,
University of Chicago
Press, 1943

Page 156
bottom, B. F. Peery, P. C.
Keenan, and I. R. Marenin,
Mt. Wilson and Las Cam-
panas Observatories

Page 157
From *An Atlas of the Spectra of the
Cooler Stars* by P. C. Keenan
and R. C. McNeil, Ohio State
University Press, 1976

Page 158
P. R. Jewell, J. C. Webber and
L. E. Snyder

Page 159
top, S. Ridgeway and J. Christou,
NOAO
bottom, Bruce Balick, NOAO

Page 160
L. H. Aller, Lick Observatory

Page 162
From *Stars and Their Spectra* by
J. B. Kaler, Cambridge Uni-
versity Press

Page 163
right, Bruce Balick
far left, Bruce Balick

Page 164
Noam Soker and Phillip Plait

Page 165
top, Richard A. Shaw
bottom, Bo Reipurth, ESO

Page 167
Jesse L. Greenstein, Palomar Observatory

Page 168
From *Voyage through the Universe:
Stars Art* by
Damon M. Hertig and Daniel
Rodriguez, Time/Life Books,
1988.

Page 169
top, David Joel/University of Chicago
bottom, Allan E. Morton

Page 170
F. Paresce, R. Jedrzejewski, Space
Telescope Science Institute,
NASA/ESA

Page 171
top, P. W. Merrill, Mt. Wilson and
Las Campanas Observatories
left, Andrew Michalitsianos
right, Space Telescope Science Institute, NASA

Page 172
Royal Observatory, Edinburgh

Page 174
left, Anglo-Australian Telescope
Board 1981
right, C. R. O'Dell and S. K.
Wong (Rice Univ.), NASA

Page 175
top, Anglo-Australian Telescope
Board 1977
bottom, Anglo-Australian Telescope Board 1984

Page 176
top, Anglo-Australian Telescope
Board 1984
bottom, ROE/AAT Board 1984

Page 177
Astronomy Magazine

Page 178
A. Maeder and G. Meynet

Page 180
From *Atlas of Representative Spectra* by Y. Yamashita, K. Nariai, and Y. Norimoto, University of Tokyo Press

Page 181
From *Atlas of Representative Spectra* by Y. Yamashita, K. Nariai, and Y. Norimoto, University of Tokyo Press

Page 182
top, NASA, IUE, Goddard Space Flight Center
left, Anglo-Australian Telescope Board 1981
right, Royal Observatory, Edinburgh

Page 183
left, Jon Morse (Univ. of Colorado) and NASA
right, Gerd Weigelt

Page 184
left, *From Atlas of Representative Spectra* by Y. Yamashita, K. Nariai, and Y. Norimoto, University of Tokyo Press
right, Anglo-Australian Telescope Board 1979

Page 185
R. Humphreys and K. Davidson

Page 187
top, Anglo-Australian Telescope Board 1987
bottom, Space Telescope Science Institute, NASA

Page 194
John Dickel, NRAO

Page 197
Frozen Star by George Greenstein, Freundlich Books

Page 198
left, Harnden/Center for Astrophysics, right, Jeff Hester and Paul Scowen (Arizona State University) and NASA

Page 199
Frozen Star by George Greenstein, Freundlich Books

Page 200
J. van Paradijs, William Herschel Telescope

Page 201
Jun Fukue

Page 202
Harvard-Smithsonian Observatory

Page 204
Anglo-Australian Telescope Board 1980

Page 206
Rudy Schild, Smithsonian Astrophysical Observatory

Page 207
NOAO

Page 208
top, Anglo-Australian Telescope Board 1987
bottom, Lick Observatory

Page 209
top, Anglo-Australian Telescope Board 1991
bottom, Rudy Schild, Smithsonian Astrophysical Observatory

Page 210
top right, Mt. Wilson and Las Campanas Observatories
upper left, NASA, COBE Science Working Group, Goddard Space Flight Center
bottom left, COBE Science Working Group, Goddard Space Flight Center

Page 211
P. A. G. Scheuer, R. A. Laing, and R. A. Perley, NRAO/AUI

Page 212
left, Geneva Observatory
upper right, Donald P. Schneider

Page 213
S. J. Maddox, W. J. Sutherland, G. P. Epstathioun, and J. Loveday, Oxford Astrophysics

Page 214
Vera Rubin

Page 219
left, Brad Whitmore, Space Telescope Science Institute, and NASA
right, Mt. Wilson and Las Campanas Observatories

Page 220
NASA, IUE, Goddard Space Flight Center

Page 221
James Wehmer

Page 222
G. Westerhout, U. S. Naval Observatory

Page 223
K. Akabane, M. Morimoto, and M. Ishiguro, Nobeyama Radio Observatory

Page 224
Thomas M. Dame

Page 225
Dr. Ian Gatley and Dr. Ron Probst/NOAO

Page 228
Steven W. Stahler; tracks by Icko Iben, Jr.

Page 229
R. Mundt and T. Ray, Max Planck Society

Page 230
top, R. Snell
bottom, Bo Reipurth/ESO
margin, F. P. Schloerb, University of Massachusetts at Amherst

Page 231
C. Burrows (Space Telescope Science Institute), the WFPC 2 Investigation Definition Team, and NASA

Pages 232–233
Alan Boss

Page 234
UK Schmidt, Royal Observatory, Edinburgh

Page 235
IPAC, part of IRA/JPL
insert, Charles Lada and Brick Young, Steward Observatory, University of Arizona

Page 238
R. Terrile/JPL

Page 239
B. Campbell, G. Walker, and S. Young

Page 242
Jim Riffle, Cloudcroft, NM

Page 244
NASA

Page 260
National Gallery, London

Further Readings

(presented in the order of the text)

BOOKS

Pannekoek, A. *A History of Astronomy.* 1961. Reprint. New York: Dover Publications, 1989.

Berry, Arthur. *A Short History of Astronomy: From Earliest Times Through the 19th Century.* 1898. Reprint. New York: Dover Publications, 1961.

Menzel, D. H. and Pasachoff, J. *A Field Guide to the Stars and Planets.* 2nd ed. Boston: Houghton-Mifflin, 1983.

Smart, W. M. *Spherical Astronomy.* 6th ed. Cambridge, England: Cambridge University Press, 1977.

Allen, R. H. *Star Names: Their Lore and Meaning.* 1899. Reprint. New York: Dover Publications, 1963.

Ridpath, Ian. *Star Tales.* New York: Universe Books, 1988.

King, Henry C. *The History of the Telescope.* 1955. Reprint. New York: Dover Publications, 1979.

Burnham, Robert R. *Burnham's Celestial Handbook: An Observer's Guide to the Universe Beyond the Solar System.* 3 vols. New York: Dover Publications, 1978.

Kaler, James B. *Stars and Their Spectra: An Introduction to the Spectral Sequence.* Cambridge, England: Cambridge University Press, 1989.

Aller, Lawrence H. *Atoms, Stars, and Nebulae.* Cambridge, Mass.: Harvard University Press, 1971.

Bok, Bart J. and Priscilla F. *The Milky Way.* Cambridge, Mass.: Harvard University Press, 1981.

Wentzel, Donat G. *The Restless Sun.* Washington, D. C.: Smithsonian Institution Press, 1989.

Marschall, Laurence A. *The Supernova Story.* New York: Plenum Publishers, 1988.

Greenstein, George. *Frozen Star.* New York: Freudlich, 1983.

Kaufmann, William J. *Black Holes and Warped Spacetime.* New York: W. H. Freeman, 1979.

Hodge, Paul W. *Galaxies.* Cambridge, Mass.: Harvard University Press, 1986.

Kaufmann, William J. *Galaxies and Quasars.* New York: W. H. Freeman, 1979.

Riordan, Michael and Schramm, David N. *The Shadows of Creation: Dark Matter and the Structure of the Universe.* New York: W. H. Freeman, 1991.

Verschuur, G. L. *Interstellar Matters.* New York: Springer-Verlag, 1989.

Cohen, Martin. *In Darkness Born: The Story of Star Formation.* Cambridge, England: Cambridge University Press, 1988.

Goldsmith, D. and Owen, T. *The Search for Life in the Universe.* Redwood City, Calif.: Benjamin-Cummings Publishers, 1980.

"A Basic Astronomy Library," a list of 109 astronomy books for adults and 11 books for children, is available from the Astronomical Society of the Pacific, 390 Ashton Avenue, San Francisco, CA 94112, as is a catalogue of slides and videotapes.

MAPS AND ATLASES

Atlas 2000, by Wil Tirion. (deep atlas to 8th magnitude and all significant nonstellar objects)

Norton's Star Atlas. (all naked-eye stars and many telescopic objects)

SC1, SC2, SC3 Star Charts, Sky Publishing. (simple charts for beginning constellation study)

Uranometria 2000, by Wil Tirion. (very deep atlas to 10th magnitude with nonstellar objects)

ASTRONOMY MAGAZINES*

Astronomy, published by Kalmbach, Waukesha, WI.

Griffith Observer, published by the Griffith Observatory, Los Angeles, CA.

Mercury, published by the Astronomical Society of the Pacific, San Francisco, CA.

Odyssey, children's magazine, published by Kalmbach.

Planetary Report, published by the Planetary Society, Pasadena, CA.

Sky and Telescope, published by Sky Publishing, Cambridge, MA.

StarDate, published by the McDonald Observatory, Austin, TX.

GENERAL MAGAZINES WITH ASTRONOMICAL CONTENT

American Scientist, published by Sigma Xi, New York, NY.

Science News, published by Science News, New York, NY.

Scientific American, published by Scientific American, Inc., New York, NY.

*List courtesy of the Astronomical Society of the Pacific.

Subject Index

Note: Names of stars follow in a separate index.

Mars, 144, 154, 237, 239
  god of war, 14
Masers, 158
Mass, 65
  of Earth, 93
  of Jupiter, 94, 150
  lower stellar limit, 130
  of Moon, 94
  of stars, 93–97, 130
  of Sun, 94
  upper stellar limit, 131
Mass accretion
  T Tauri stars, 227
  white dwarfs, 170
Mass loss, 157–159
  from giants, 158
  from the Sun, 116
  from O stars, 182
  from T Tauri stars, 227
Mass–luminosity relation, 95, 96
Mass transfer, 169–171
Maunder minimum, 118
  for stars, 136
Maury, Antonia, 77, 78, 87, 180
Max Planck Institute 100-meter
  telescope, 54
McNaught, Robert, 191
Medusa, 16
Mercury, 145–146, 154, 159, 236–
  237
Messier, Charles, 175
Metallic-line stars, 136
Metals, 82
  in early Galaxy, 216
  globular cluster deficiency in,
    103
  subdwarf deficiency in, 90
Meteorites, 237
Michelson, A. A., 59
Microscopium, 17
Midnight Sun, 12, 13
Milky Way, 18–20, 242
  mythology of, 18, 260
Milky Way Galaxy. See Galaxy, the
Millisecond pulsars, 200
Mira variables. See Long-period
  variables
Missing mass. See Dark matter
MKK system, 91
Monoceros OB 1, 175
Moon
  age, 107
  distance, 32
  formation, 236–237
  mass, 94
  observatory scheduling, 53

Morgan, W. W., 91
Morton salt mine neutrino detec-
  tor, 191
Mount Wilson
  interferometer, 59
  100-inch telescope, 46
Moving cluster method, 92
Muon, 130

N stars, 85
Nadir, 7
Naked T Tauri stars, 228
Nebulium, 161, 163
Neptune, 159, 236
Neutrino telescopes, 127, 128, 129,
  191
Neutrinos, 121, 127–130
  and Big Bang, 205
  and mass of Universe, 215
  masses of, 130
  and solar cycle, 129
  from Supernova 1987A, 191
  from supernovae, 190
Neutron, 80, 121
Neutron stars, 196–200
  density of, 197
Newton, Isaac, 44, 65, 75
Newtonian focus, 45
Newton's laws of motion, 65
NGC 147, 207
NGC 185, 207
NGC 205, 207
NGC 752, 150
NGC 801, 214
NGC 1365, 209
NGC 1613, 207
NGC 2359, 184
NGC 2362, 150
NGC 2440, 160
NGC 3293, 149
NGC 4038, 219
NGC 4039, 219
NGC 4565, 20
NGC 6164-5, 182
NGC 6822, 207
NGC 6826, 164
NGC 7000. See North America
  Nebula
NGC 7009, 159
NGC 7662, 163
North America Nebula, 176
Northern Cross, 19
Northern lights. See Aurora
Nova Herculis 1934 shell, 170
Nova shells, 170
Novae, 170

Nuclear force. See Strong force
Nucleosynthesis, 150
  in supernovae, 191
Nutation, 31

O stars and the interstellar me-
  dium, 226
OB associations, 100
  and supergiants, 177
Objective prism spectrogram, 77
Objective prism spectrograph, 78
Observational selection, 93
Observatories, 48
Oe stars, 180
Of stars, 180, 181
OH, interstellar, 222
OH/IR stars, 158
Omega, 214
Omega Centauri, 102
Oort, Jan, 237
Oort comet cloud, 237
Opacity, 124
  of Sun, 107
Open clusters, 101
  ages, 150
  discovery of dust, 219
  disruption of, 150
  HR diagrams of, 93, 150
Ophiuchus, 20
Ophiuchus dark cloud, 235
Optics, principles of, 37
Orbits, 64–67
  of binary stars, 94
  electron, 80
  of hydrogen atom, 83
Organic molecules, interstellar, 222
Orion, 14, 15
  in infrared, 225
Orion Nebula, 173, 174
Oxygen burning, 188

P Cygni lines, 180
Palomar
  sky survey, 50
  200-inch telescope, 46
Parallax, 9
  of planets, 66
  of stars, 68
Parkes radio telescope, 54
Parsec, 68
Paschen lines, 83
Pauli, Wolfgang, 121, 146
Payne-Gaposchkin, Cecilia, 121
PC 1158+4635, 212
Perfect gas law, 123, 146

Periodic table, 81
Perseus, 16
Phase space, 146
Photoelectric photometer, 50
Photography, astronomical, 49
Photon, 36
Photosphere, 107
Pickering, E. C., 77
Planck time, 205
Planck's constant, 36, 121
Planet formation, 236–239
Planetary nebulae, 159–165
  compositions, 163
  evolution of central stars of, 164
  spectrum, 160
Planetesimals, 236
Pleiades, 91, 101
  age, 149
  HR diagram, 93, 141
Plough, 14
Polarized radiation, 195
Polycyclic aromatic hydrocarbons,
  224
Population I, 101
  creation of, 217
Population II, 101
  and elliptical galaxies, 208
Population III, 216
Positron, 121
p–p reaction. See Proton–proton
  reaction
Praesepe cluster, 150
Precession, 30
Prime meridian, 23
Prism, 38
Project Ozma, 240
Prominences, 113
  eruptive, 115
Proper motion, 67
Proton, 80
Proton–proton reaction, 121, 122
  alternate cycles, 128
Protostars, 233
Ptolemaic theory, 32
Ptolomy, 13
Pulsars, 196–200. See also Neutron
  stars
  cause of, 198
  glitches of, 197
  internal construction of, 199
  millisecond, 200
  planet of, 239
  slowing of rotation, 197
Puppis, 18
Purcell, Edward, 220
Pythagoras, 6, 31

Star Index

Stars are presented alphabetically by proper or catalogue name, then by constellation in the following order: Greek letter name, Roman letter name, Flamsteed number, variable star name in discovery order, OB association name. Spectra are indicated by asterisks.